Reasonable Books = Reasonable Prices

Irina Kleyman

AF877845

Exceptional for Home Schooling

Math Help Grade 6

First Edition

Reasonable Books = Reasonable Prices

Irina Kleyman

Exceptional for Home Schooling

Math Help
Grade 6

I dedicate this book to my children and grandchildren.

Acknowledgments
Special thanks to Michael G LaFratta, Vladimir Kleyman,
who has assisted in bringing all 10 books to market.

Author's note: In the many years that I've taught mathematics, I have been informed by many of my students that my methods for teaching have helped them tremendously in learning math. I have written these books in order to continue to help others learn math effectively.

Math Help Grade 1
Math Help Grade 2
Math Help Grade 3
Math Help Grade 4
Math Help Grade 5
Math Help Grade 6
Math Help Grade 7
Math Help Grade 8
Math Help for College (Summary Grades 1 – 5)
Math Help for University (Basic, Algebra, Geometry)

ISBN: 978-1-7349699-5-5 Paperback

Syllabus Math 5, 6, 7, 8

Grade 5 # 1–5, Grade 6 # 6–10, Grade 7 # 11–15, Grade 8 # 16–20

	Grades	Topics: Advanced Problems (A), Pages (p)
#	Chapters	Basic Problems (B), Chapter (C), Lesson (L)
1	5-1	Four–digit numbers (+, -) (B)
2	5-2	Multiplying numbers (B)
3	5-3	Dividing numbers (B)
4	5-4	Four operations with numbers (B)
5	5-5	Finding unknowns (A-5)
	Grade 5	Final Exam 5 B, C – 5, L 16 #1-10 all, p.150
6	6-1	Many-digit numbers (B)
7	6-2	Multiplying numbers (B)
8	6-3	Dividing numbers (B)
9	6-4	Four operations with numbers (B)
10	6-5	Finding unknowns (A-6)
	Grade 6	Final Exam 6 B, C 5, L 16 #1-10 all, p.196
11	7-1	Whole numbers (B)
12	7-2	Properties of whole numbers (B)
13	7-3	Decimals (B)
14	7-4	Fractions and mixed numbers (B)
15	7-5	Converting fractions and decimals (B)
	Grade 7	Final Exam 7 B, C – 5, L 12 #1-10 all, p.209-210
16	8-1	Rational numbers (B)
17	8-2	Equations (B)
18	8-3	Geometry (B)
19	8-4	Word Problems (B)
20	8-5	Summary of Basic Math (B)
21	8-6	Extra Review (B)
	Grade 8	Final Exam 8 B, C 5, L 14 #1-10 all, p.281 -282

Middle School can start from 5th, 6th, or 7th grade. Do not worry about school. Choose your grade, work hard and you have good knowledge.
Syllabus includes topics, what students need to know.
Unknowns are numbers, which we need to find.

Grade 6
Chapter 1

Many-digit Numbers

www.mathbooksbyirina.com

Parents, Tutors, Students Middle Schools, High Schools, Colleges, University can help primary school students. Please read slowly math books to primary school students.

Grade 6 Chapter 1 Contents
Many-digit numbers
Lesson 1 – 14
Basic Problems

Work with lesson 1, read rules and examples, 2) Solve the problems from contents, 3) Check answers in the end of chapter 1, 4) Work with the next lessons 2, 3, 4, 5, …, 5) Solve Test 1 A, 6) Grading Scale: A =10 points, B=9points, C=8 points, D=7 points, F=0-6 points, 7) If you solve 10 problems correctly, you get an A, 8) If have an A, you can move on to next chapter, 9) If your grade is lower than an A, work with the mistakes, then solve Test 1B, 10) When you finish all of the chapters, solve Final Exam in last chapter.

Grade 6 Chapter 1 Lesson 1

Four-digit numbers

Four-digit numbers content four digits: 1000, 1001 … 9999
Least four-digit number: 1000. Greatest four-digit number: 9999
Digit counts from right to the left: ones, tens, hundreds, thousands
2468 = 2000 + 400 + 60 + 8
One-digit numbers always less than two, three, four-digit numbers
Four-digit numbers always greater than one, two, three-digit numbers
To compare numbers, compare thousands, hundreds, tens, and ones:

$1234 < 1239 \quad (4 < 9) \qquad 1238 > 1230 \; (8 > 0) \qquad 1234 = 1234$

Properties of Addition and Subtraction

Two replace addends the sum does not change $A + B = B + A$
2000 + 3000 = 5000 3000 + 2000 = 5000
The sum of a number and zero is the same number
5000 + 0 = 5000 0 + 5000 = 5000 $A + 0 = 0 + A = A$
Difference any number and zero is the same number
5000 – 0 = 5000 $A - 0 = A$
The difference two the same number is zero
5000 – 5000 = 0 $A - A = 0$

10 = 1 + 9 = 2 + 8 = 3 + 7 = 4 + 6 = 5 + 5
11 = 2 + 9 = 3 + 8 = 4 + 7 = 5 + 6
12 = 3 + 9 = 4 + 8 = 5 + 7 = 6 + 6
13 = 4 + 9 = 5 + 8 = 6 + 7
14 = 5 + 9 = 6 + 8 = 7 + 7
15 = 6 + 9 = 7 + 8
16 = 7 + 9 = 8 + 8
17 = 8 + 9
18 = 9 + 9

Adding four-digit numbers:
Add the ones. Add the tens. Add the hundreds. Add the thousands.
Subtracting four- digit numbers:
Subtract the ones. Subtract the tens. Subtract the hundreds. Subtract the thousands.

		1 1 1	8 13 14 16
5678	9678	2686	9 4 5 6
+ 4321	- 5432	+ 1868	- 3 9 8 7
9999	4246	4554	5 4 6 9

Grade 6 Chapter 1 Lesson 1
Review 1 A
Add numbers

	a	b	c	d	e
# 1	1919 +1191	2828 +1282	3737 +1373	4646 +1464	5959 + 1595
# 2	2929 +1292	3838 +1383	4747 +1474	5656 +1565	6868 +1686
# 3	3939 +1393	4848 +1484	5757 +1575	6969 +1696	7979 +1797
# 4	4949 +1494	5858 +1585	6767 +1676	7878 +1787	8989 +898

Review 1 B
Subtract numbers

	a	b	c	d	e
# 5	3110 -1919	4110 -2828	5110 - 3737	6110 -4646	7554 -5959
# 6	4221 -2929	5221 -3838	6221 -4747	7221 -5656	8554 -6868
# 7	5332 -3939	6332 -4848	7332 -5757	8665 -6969	9776 -7979
# 8	6443 -4949	7443 -5858	8443 -6767	9665 -7878	9887 -8989

Grade 6 Chapter 1 Lesson 2

Five-digit Numbers

Five-digit number content five digits: 10000, 10001 … 99999
Least five-digit number: 10000.
Greatest five-digit number: 99999
Digit counts from right to the left: ones, tens, hundreds, thousands, ten thousand. Twelve thousand four hundred sixty-eight:
12468 = 10000 + 2000 + 400 + 60 + 8
One-digit numbers always less than two, three, four, five-digit numbers
Five-digit numbers always greater than one, two, three, four-digit numbers
To compare number, compare ten thousand, thousands, hundreds, tens, ones
12345 = 12345   12345 < 12349 (5 < 9)   12348 > 12340 (8 > 0)

Properties of addition and subtraction

To replace addends the sum does not change: A + B = B + A
20000 + 30000 = 50000   30000 + 20000 = 50000
Sum any number and zero is the same number
50000 + 0 = 50000   0 + 50000 = 50000   A + 0 = 0 +A = A
Difference any number and zero is the same number
50000 – 0 = 50000   A – 0 = A
The difference two the same number is zero
50000 – 50000 = 0   A – A = 0

Adding five-digit numbers:
Add the ones. Add the tens. Add hundreds.
Add the thousands. Add the ten thousand.

Subtracting five-digit numbers: Subtract the ones. Subtract the tens.
Subtract the hundreds. Subtract the thousands. Subtract the ten thousand.

		1 1 1	8 13 14 16
56780	96780	26860	9 4 5 6 0
+ 43210	- 54320	+ 18680	- 3 9 8 7 0
99990	42460	45540	5 4 6 90

Grade 6 Chapter 1 Lesson 2
Review 1 A
Add numbers

	a	b	c	d	e
# 1	19190	28280	37370	46460	59590
	+11910	+12820	+13730	+1464 0	+ 15950
# 2	29290	38380	47470	56560	68680
	+12920	+13830	+14740	+15650	+16860
# 3	39390	48480	57570	69690	79790
	+13930	+14840	+15750	+16960	+17970
# 4	49490	58580	67670	78780	89890
	+14940	+15850	+16760	+17870	+8980

Review 1 B
Subtract numbers

	a	b	c	d	e
# 5	31100	41100	51100	61100	75540
	-19190	-28280	- 37370	-46460	-59590
# 6	42210	52210	62210	72210	85540
	-29290	-38380	-47470	-56560	-68680
# 7	53320	63320	73320	86650	97760
	-39390	-48480	-57570	-69690	-79790
# 8	64430	74430	84430	96650	98870
	-49490	-58580	-67670	-78780	-89890

Grade 6 Chapter 1 Lesson 3
Six-digit Numbers

Six-digit number content six digits 100000, 100001, … 999999
Least six-digit number is 100000; greatest six-digit number is 999999
Digit counts from right to the left - ones, tens, hundreds, thousands, ten thousand, hundred thousand: 123456 = 100000+20000+3000+400+50+6

Compare six-digit numbers

One-digit number always less than two, three, four, five, six- digit number
Six-digit number always greater than five, four, three, two, one- digit number.
To compare number, compare hundred thousand, ten thousand, thousands, hundreds, tens, ones

123456 = 123456

123456 < 123459 (6 < 9) **123458 > 123450 (8 > 0)**

Properties six-digit numbers: Letters can represent any numbers

To replace addends the sum does not change: A + B = B + A
200000 + 300000 = 500000 **300000 + 200000 = 500000**

The sum of a number and zero is the same number: A + 0 = 0 + A = A
500000 + 0 = 500000 **0 + 500000 = 500000**

Difference any number and zero is the same number
500000 – 0 = 500000 **A – 0 = A**

The difference two the same number is zero
500000 – 500000 = 0 **A – A = 0**

Addition six-digit numbers: Add the ones. Add the tens. Add the hundreds. Add the thousands. Add the ten thousand. Add the hundred thousand
Subtraction six-digit numbers: Subtract the ones. Subtract the tens. Subtract the hundreds. Subtract the thousands. Subtract the ten-thousands. Subtract the hundred-thousand

		1 1 1 1	8 17 16 15 14 13
654321	777777	456789	9 8 7 6 5 3
+123456	-123456	+345678	- 7 8 9 7 8 9
777777	654321	802467	1 9 7 8 6 4

Grade 6 Chapter 1 Lesson 3
Review 1 A
Add six-digit numbers

	a	b	c	d	e
1)	191900	282800	373700	464600	595900
	+119100	+128200	+137300	+146400	+159500
2)	292900	383800	474700	565600	686800
	+129200	+138300	+147400	+156500	+168600
3)	393900	484800	575700	696900	797900
	+139300	+148400	+157500	+169600	+179700
4)	494900	585800	676700	787800	898900
	+149400	+158500	+167600	+178700	+89800

Review 1 B
Subtract six-digit numbers

	a	b	c	d	e
5)	311000	411000	511000	611000	755400
	-191900	-282800	- 373700	-464600	-595900
6)	422100	522100	622100	722100	855400
	-292900	-383800	-474700	-565600	-686800
7)	533200	633200	733200	866500	977600
	-393900	-484800	-575700	-696900	-797900
8)	644300	744300	844300	966500	988700
	-494900	-585800	-676700	-787800	-898900

Grade 6 Chapter 1 Lesson 4

Seven-digit numbers

Seven-digit number contain seven digits: 1,000,000 … 9,999,999
The name of seven-digit number read from right to left are millions, hundred-thousands, ten-thousands, thousands, hundreds, tens, ones.

Standard form of numbers is 123, 1234, 12345, 123456.
The expanded form of number
1,234,567 = 1,000,000 + 200,000 + 30,000 + 4,000 + 500 + 60 + 7 =
1,000,000 + 2×100,000 + 3×10,000 + 4×1,000 + 5×100 + 6×10 + 7

Addition seven-digit numbers: Add the ones, add the tens, add the hundreds, and add the thousands, Add the ten-thousands, add the hundred-thousands, and add the millions.

Subtraction seven-digit numbers: Subtract the ones, subtract the tens, subtract the hundreds, Subtract the thousands, subtract the ten-thousands Subtract the hundred-thousands, subtract the millions.

```
      1  1           10 12
   1020304         6081012
 + 5060708       - 1020304
   6081012         5060708
```

Many-digit Numbers

Third new period has nine digit and called million 123, 456, 789
The place value of the third period read from right to left are hundred-millions is 1, ten-millions is 2, and millions is 3, hundred-thousand is 4, ten-thousands is 5, thousands is 6, hundreds is 7, tens is 8, and ones is 9

Roman Numbers

I means 1	V means 5	X means 10
II means 1+1=2	III means 1+1+1=3	IV means 5–1= 4
VI means 5+1=6	VII means 5+1+1=7	VIII means 5+1+1+1=8
IX means 10-1=9	XI means 10+1=11	XII means 10+2 = 12
XIV means 10+4=14	XIX means 10+9=19	XXV=10+10+5 = 25

Grade 6 Chapter 1 Lesson 4

Review 1 A

Add numbers

	a	b	c	d
1)	1919000 +1191000	2828000 +1282000	3737000 +1373000	4646000 +1464000
2)	2929000 +1292000	3838000 +1383000	4747000 +1474000	5656000 +1565000
3)	3939000 +1393000	4848000 +1484000	5757000 +1575000	6969000 +1696000
4)	4949000 +1494000	5858000 +1585000	6767000 +1676000	7878000 +1787000
5)	5959000 +1590000	6868000 +1686000	7979000 +1797000	8989000 +898000

Review 1 B

Subtract numbers

	a	b	c	d
6)	3110000 -1919000	4110000 -2828000	5110000 -3737000	6110000 -4646000
7)	4221000 -2929000	5221000 -3838000	6221000 -4747000	7221000 -5656000
8)	5332000 -3939000	6332000 -4848000	7332000 -5757000	8665000 -6969000
9)	6443000 -4949000	7443000 -5858000	8443000 -6767000	9665000 -7878000
10)	7554000 - 5959000	8554000 -6868000	9776000 -7979000	9887000 -8989000

Grade 6 Chapter 1 Lesson 5

Multiplication Numbers 1

Third operation with numbers is multiplication $\times$ or $\cdot$
Multiplication is the operation of repeated addition
$2\times 3 = 2 + 2 + 2 = 6$ or $2 \cdot 3 = 2 + 2 + 2 = 6$

First number in multiplication called the factor 2
Second number in multiplication called the factor 3
Result of multiplication called the product 6
Factor $\times$ Factor = Product $2 \times 3 = 6$

Properties of multiplication
Two numbers can multiply in any order
The product will be the same $AB = BA$
$2 \times 3 = 2 + 2 + 2 = 6$ $3 \times 2 = 3 + 3 = 6$
The product of a number and zero is zero
$A \times 0 = 0 \times A = 0$ $5 \times 0 = 0$ $0 \times 5 = 0$
The product of a number and one is the same number
$A \times 1 = 1 \times A = A$ $5 \times 1 = 5$ $1 \times 5 = 5$

Factor $\times$ Factor = Product (P) $2\times3 = 6$ (P)

1. First way

$$\begin{array}{r} 25 \\ \underline{\times 6} \\ 30 \\ \underline{+\,120} \\ 150 \text{ (P)} \end{array}$$

1) $25 = 20 + 5$
2) $6 \times 5 = 30$
3) $6 \times 20 = 120$
4) $30 + 120 = 150$

2. Second way

$$\begin{array}{lr} \text{Top} & 3\ \ \\ & 25 \\ & \underline{\times 6} \\ & 150 \text{ (P)} \end{array}$$

1) $6 \times 5 = 30$
2) Put 0 below 6 (P)
3) Put 3 above 2 (T)
4) $6 \times 2 = 12$
5) $12 + 3 = 15$
6) Put 15 ahead 0 (P)

Memorize Rules and Table of Multiplication

$A + 0 = A$ $0 + A = A$ $A - 0 = A$ $A - A = 0$

$A \times 1 = A$ $1 \times A = A$ $A \div 1 = A$ $A \div A = 1$

$A \times 0 = 0$ $0 \times A = 0$ $0 \div A = 0$ $A \div 0$ None

Table of Multiplication and Division

Number 5 from the left column multiply by number 4 from first row to get answer 20

×÷	1	2	3	<u>4</u>	5	6	7	8	9	10
1	1	2	3	4	5	6	7	8	9	10
2	2	4	6	8	10	12	14	16	18	20
3	3	6	9	12	15	18	21	24	27	30
4	4	8	12	16	20	24	28	32	36	40
<u>5</u>	5	10	15	20	25	30	35	40	45	50
6	6	12	18	24	30	36	42	48	54	60
7	7	14	21	28	35	42	49	56	63	70
8	8	16	24	32	40	48	56	64	72	80
9	9	18	27	36	45	54	63	72	81	90
10	10	20	30	40	50	60	70	80	90	100

Grade 6 Chapter 1 Lesson 6
Multiplication two-digit numbers 2

Factor × Factor = Product $2 \times 3 = 6$ (Product)

First way

$$\begin{array}{r} 69 \\ \times\underline{52} \\ 138\ (P1) \\ +\underline{3450\ (P2)} \\ 3588\ (P) \end{array}$$

1) 52=50+2
2) $69 \times 2 = 138$ (P1)
3) $69 \times 50 = 3450$ (P 2)
4) $138 + 3450 = 3588$

Second way

Top 4
Top 1

$$\begin{array}{r} 69 \\ \times\underline{52} \\ 138\ (P1) \\ +\underline{3450\ (P2)} \\ 3588\ (P) \end{array}$$

1) $2 \times 9 = 18$
2) Put 8 below 2 (P1)
3) Put 1 above 6 (top)
4) $2 \times 6 = 12$
5) $12 + 1 = 13$
6) Put 13 ahead 8 (P1)
7) $5 \times 9 = 45$
8) Put 5 below 3 (P2)
9) Put 4 above 1 (top)
10) $5 \times 6 = 30$
11) $30 + 4 = 34$
12) Put 34 ahead 5 (P2)

Review 1 A
Multiply

	a	b	c	d	e
# 1	50× 55	51× 55	52× 55	53× 55	54× 55
# 2	55× 55	56× 55	57× 55	58× 55	59× 55
# 3	50× 56	51× 56	52× 56	53× 56	54× 56
# 4	55× 56	56× 56	57× 56	58× 56	59× 56
# 5	50× 57	51× 57	52× 57	53× 57	54× 57

Review 1B

	a	b	c	d	e
# 6	60×55	61×55	62×55	63×55	64×55
# 7	65×55	66×55	67×55	68×55	69×55
# 8	60×56	61×56	62×56	63×56	64×56
# 9	65×56	66×56	67×56	68×56	69×56
# 10	60×57	61×57	62×57	63×57	64×57

Grade 6 Chapter 1 Lesson 6
Multiplication Numbers 2

1. First way

```
  525
 × 6
   30
  120
+3000
 3150
```

1) 525 = 500 + 20 + 5
2) 6 × 5 = 30
3) 6 × 20 = 120
4) 6 × 500 = 3000

2. Second way

```
Top 1 3
    525
   × 6
   3150
```

1) 6×5 = 30
2) Put 0 below 6 (P)
3) Put 3 above 2 (top)
4) 6 × 2 = 12
5) 12 + 3 = 15
6) Put 5 ahead 0 (P)
7) Put 1 ahead 3 (top)
8) 6 × 5 = 30
9) 30 + 1 = 31
10) Put 31 ahead 5 (P)

Review 2 A Multiply

	a	b	c	d
# 1	534×5	567×5	656×6	678×6
# 2	745×7	789×7	856×8	867×8
# 3	967×9	978×9	543×5	576×5
# 4	675×6	687×6	754×7	798×7
# 5	865×8	876×8	976×9	985×9

Review 2 B Multiply

	a	b	c	d
# 1	546 × 5	658 × 5	645 × 6	689 × 6
# 2	756 × 7	768 × 7	867 × 8	879 × 8
# 3	978 × 9	987 × 9	564 × 5	685 × 5
# 4	654 × 6	698 × 6	764 × 7	786 × 7
# 5	876 × 8	897 × 8	967 × 9	976 × 9

Grade 6 Chapter 1 Lesson 7

Multiplication Numbers 3

Factor × Factor = Product (P) $2 \times 3 = 6$ **(P)**

Example 1

Top 3 2	1) $5 \times 4 = 2\underline{0}$	7) $5 \times 7 = 3\underline{5}$
6704	2) Put $\underline{0}$ below 5 (P)	8) Put $\underline{5}$ ahead 2 (P)
$\underline{\times\ \ 5}$	3) Put 2 above 0 (top)	9) Put 3 above 6 (top)
33520 (P)	4) $5 \times 0 = 0$	10) $5 \times 6 = 30$
	5) $2 + 0 = \underline{2}$	11) $30 + 3 = \underline{33}$
	6) Put $\underline{2}$ ahead 0 (P)	12) Put $\underline{33}$ ahead 5 (P)

Example 2

Top 2 3	1) $5 \times 6 = 3\underline{0}$	7) $5 \times 4 = 2\underline{0}$
7406	2) Put $\underline{0}$ below 5 (P)	8) Put $\underline{0}$ ahead 3 (P)
$\underline{\times\ \ 5}$	3) Put 3 above 0 (top)	9) Put 2 above 7 (top)
37030 (P)	4) $5 \times 0 = 0$	10) $5 \times 7 = 35$
	5) $3 + 0 = \underline{3}$	11) $35 + 2 = \underline{37}$
	6) Put $\underline{3}$ ahead 0 (P)	12) Put $\underline{37}$ ahead 0 (P)

Example 3

Top 2 3	1) $5 \times 6 = 3\underline{0}$	7) $5 \times 4 = 2\underline{0}$
8406	2) Put $\underline{0}$ below 5 (P)	8) Put $\underline{0}$ in ahead 3 (P)
$\underline{\times\ \ 5}$	3) Put 3 above 0 (top)	9) Put 2 above 8 (top)
42030 (P)	4) $5 \times 0 = 0$	10) $5 \times 8 = 40$
	5) $3 + 0 = \underline{3}$	11) $40 + 2 = \underline{42}$
	6) Put $\underline{3}$ ahead 0 (P)	12) Put $\underline{42}$ ahead 0 (P)

Example 4

Top 2 3	1) $5 \times 6 = 3\underline{0}$	7) $5 \times 4 = 2\underline{0}$
9406	2) Put $\underline{0}$ below 5 (P)	8) Put $\underline{0}$ ahead 3 (P)
$\underline{\times\ \ 5}$	3) Put 3 above 0 (top)	9) Put 2 above 9 (top)
47030 (P)	4) $5 \times 0 = 0$	10) $5 \times 9 = 45$
	5) $3 + 0 = \underline{3}$	11) $45 + 2 = \underline{47}$
	6) Put $\underline{3}$ ahead 0 (P)	12) Put $\underline{47}$ ahead 0 (P)

Grade 6 Chapter 1 Lesson 7

Multiplication Numbers 3

Example 5

$$\begin{array}{r} {\scriptstyle 2} \\ {\scriptstyle 1} \\ 803 \\ \underline{\times\ 75} \\ 4015 \\ \underline{+56210} \\ 60225 \end{array}$$

1) $75 = 70 + 5$

2) $$\begin{array}{r} {\scriptstyle 1} \\ 803 \\ \underline{\times\ 5} \\ 4015 \end{array}$$

3) $$\begin{array}{r} {\scriptstyle 2} \\ 803 \\ \underline{\times\ 70} \\ 56210 \end{array}$$

Example 6

$$\begin{array}{r} {\scriptstyle 3} \\ {\scriptstyle 2} \\ 904 \\ \underline{\times\ 85} \\ 4520 \\ \underline{+72320} \\ 76840 \end{array}$$

1) $85 = 80 + 5$

2) $$\begin{array}{r} {\scriptstyle 2} \\ 904 \\ \underline{\times\ 5} \\ 4520 \end{array}$$

3) $$\begin{array}{r} {\scriptstyle 3} \\ 904 \\ \underline{\times\ 80} \\ 72320 \end{array}$$

Example 7

$$\begin{array}{rl} 405 & \\ \underline{\times 106} & \\ 2430 & \text{(P1)} \\ \underline{+40500} & \text{(P2)} \\ 42930 & \text{(P)} \end{array}$$

1) $106 = 100 + 6$

2) $$\begin{array}{rl} 405 & \\ \underline{\times\ 6} & \\ 2430 & \text{(P1)} \end{array}$$

3) $$\begin{array}{rl} 405 & \\ \underline{\times 100} & \\ 40500 & \text{(P2)} \end{array}$$

Example 8

$$\begin{array}{r} 704 \\ \underline{\times 205} \\ 3520 \\ \underline{+140800} \\ 144320 \end{array}$$

1) $205 = 200 + 5$

2) $$\begin{array}{r} 704 \\ \underline{\times\ 5} \\ 3520 \end{array}$$

3) $$\begin{array}{r} 704 \\ \underline{\times 200} \\ 140800 \end{array}$$

Grade 6 Chapter 1 Lesson 8

Division Numbers 1

Fourth operation with numbers is division ÷ or :

Division means separated number by some equal groups

First number in division called the dividend 6
Second number in division called the divisor 2
Result of division called the quotient 3
Dividend ÷ divisor = quotient $6 \div 2 = 3$

Properties of Division

Any number divided by one is the same number $5 \div 1 = 5$
Any number divided by the itself is one $5 \div 5 = 1$
Zero divided by any number is zero $0 \div 5 = 0$
Division by zero is not allowed $5 \div 0$ $A \div 0$ **none**

Formulas: 1) $A \div 1 = A$ 2) $A \div A = 1$
3) $0 \div A = 0$ 4) $A \div 0$ **none**

To check multiplication use division
$2 \times 3 = 6$ Check $6 \div 2 = 3$ or $6 \div 3 = 2$
To check division use multiplication and division
$6 \div 2 = 3$ Check $2 \times 3 = 6$ or $6 \div 3 = 2$

Division without remainder R = 0

```
          5  - Quotient
Divisor 7| 35 – Dividend
          -35
            0 – Remainder
```

Division with remainder R > 0

```
          5 - Quotient
Divisor 7|36 – Dividend
          -35
            1 – Remainder
```

Grade 6 Chapter 1 Lesson 8

Addition and Subtraction of Numbers

Complete the table 1

5 + 0 = 5 0 + 5 = 5 5 – 0 = 5 5 – 5 = 0

Number 5 from the left column add number 4 from first row to get answer 9

	+, -	0		2		4		6		8	
#1	1										10
#2										10	
#3	3								10		
#4								10			
#5	5					9	10				
#6						10					
#7	7				10						
#8				10							
#9	9		10								
#10		10									

Complete the table 2

Multiplication and Division of Numbers

$5 \times 1 = 5$ $1 \times 5 = 5$ $5 \times 0 = 0$ $0 \times 5 = 0$

$5 \div 1 = 5$ $5 \div 5 = 1$ $0 \div 5 = 0$ $5 \div 0$ None

Number 5 from the left column multiply by number 4 from first row to get answer 20

	× ÷		2		4		6		8	
#11	1									9
#12									16	
#13	3							21		
#14							24			
#15	5				20	25				
#16					24					
#17	7			21						
#18			16							
#19	9	9								
#20	10									

Grade 6 Chapter 1 Lesson 9
Division Numbers 2

Example 1   5400 – dividend   8 – divisor   685 – quotient

Every time bring down just one digit

675	1) $8 \times \underline{6} = 48$	Put 6 in quotient
8 | 5400	2) $54 - 48 = 6$	
-48	3) Bring down the 8	
60	4) $8 \times \underline{7} = 56$	Put 7 in quotient
- 56	5) $60 - 56 = 4$	
40	6) Bring down the 0	
-40	7) $8 \times \underline{5} = 40$	Put 5 in quotient
0	8) $40 - 40 = 0$	

Example 2   7800 – dividend   8 – divisor   975 – quotient

Every time bring down just one digit

975	1) $8 \times \underline{9} = 72$	Put 9 in quotient
8 | 7800	2) $78 - 72 = 6$	
-72	3) Bring down the 0	
60	4) $8 \times \underline{7} = 56$	Put 7 in quotient
- 64	5) $60 - 56 = 4$	
40	6) Bring down the 0	
-40	7) $8 \times \underline{5} = 40$	Put 5 in quotient
0	8) $40 - 40 = 0$	

Review 1 A   Divide numbers

	a	b		a	b
# 1	2730 ÷ 5	2670 ÷ 5	# 6	5383 ÷ 7	5523 ÷ 7
# 2	2890 ÷ 5	2835 ÷ 5	# 7	6936 ÷ 8	6672 ÷ 8
# 3	3870 ÷ 6	3936 ÷ 6	# 8	7024 ÷ 8	6848 ÷ 8
# 4	4134 ÷ 6	4068 ÷ 6	# 9	8802 ÷ 9	8703 ÷ 9
# 5	5292 ÷ 7	5215 ÷ 7	# 10	8883 ÷ 9	8874 ÷ 9

Review 1 B   Divide numbers

	a	b		a	b
# 1	2820 ÷ 5	2715 ÷ 5	# 6	5572 ÷ 7	5586 ÷ 7
# 2	2935 ÷ 5	2880 ÷ 5	# 7	7000 ÷ 8	6768 ÷ 8
# 3	3924 ÷ 6	4050 ÷ 6	# 8	7168 ÷ 8	6920 ÷ 8
# 4	4188 ÷ 6	4122 ÷ 6	# 9	8703 ÷ 9	8784 ÷ 9
# 5	5355 ÷ 7	5278 ÷ 7	# 10	8802 ÷ 9	8901 ÷ 9

Grade 6 Chapter 1 Lesson 9
Division Numbers 2

2540 – dividend   5 – divisor   508 – quotient

Example 1   **Every time bring down just one digit**

```
   508
5 |2540
  -25
   04
   -0
    40
   -40
     0
```

1) $5 \times \underline{5} = 25$ **Put 5 in quotient**
2) $25 - 25 = 0$
3) **Bring down the 4**
4) $5 \times \underline{0} = 0$ **Put 0 in quotient**
5) $4 - 0 = 4$
6) **Bring down the 0**
7) $5 \times \underline{8} = 40$ **Put 8 in quotient**
8) $40 - 40 = 0$

Example 2   **3726 – dividend   54 – divisor   69 quotient**

Every time bring down just one digit

$37 < 54$ $54 \times 6 = 324$ $54 \times 7 = 378$ $378 > 372$

```
     69
54 | 3726
    -324
      486
     -486
        0
```

1) $54 \times 6 = 324$
2) $372 - 324 = 48$
3) **Bring down the 6**
4) $54 \times 9 = 486$
5) $486 - 486 = 0$

or

```
3726 |_54
-324   69
 486
-486
   0
```

Example 3   **Every time bring down just one digit**

```
   7608
5 |38040
  -35
   30
  -30
    04
    -0
     40
    -40
      0
```

1) $5 \times \underline{7} = 35$ **Put 7 in quotient**
2) $38 - 35 = 3$
3) **Bring down the 0**
4) $5 \times \underline{6} = 30$ **Put 6 in quotient**
5) $30 - 30 = 0$
6) **Bring down the 4**
7) $5 \times \underline{0} = 0$ **Put 0 in quotient**
8) $4 - 0 = 4$
9) **Bring down the 0**
10) $5 \times \underline{8} = 40$ **Put 8 in quotient**
11) $40 - 40 = 0$

Grade 6 Chapter 1 Lesson 10
Division Numbers 3

Example 1

Every time bring down just one digit

```
     507
86 | 43602
   -430
      60
      -0
     602
   - 602
       0
```

1) $86 \times \underline{5} = 430$ Put 5 in quotient
2) $436 - 430 = 6$
3) Bring down the 0
4) $86 \times \underline{0} = 0$ Put 0 in quotient
5) $60 - 0 = 60$
6) Bring down the 2
7) $86 \times \underline{7} = 602$ Put 7 in quotient
8) $602 - 602 = 0$

Example 2

6÷2=3 dividend 6 divisors 2 quotients 3(q)

Every time bring down just one digit

```
       708 (q)
905 | 640740
    -6335
       7240
      -7240
```

1) $905 \times \underline{7} = 6335$ Put 7 in quotient (q)
2) $6407 - 6335 = 72$
3) Bring down the 4
4) $905 \times \underline{0} = 0$ Put 0 in quotient (q)
5) $724 - 0 = 724$
6) Bring down the 0
7) $905 \times \underline{8} = 7240$ Put 8 in quotient (q)
8) $7240 - 7240 = 0$

Example 3

Every time bring down just one digit

```
       807 (q)
506 | 408342
    -4048
      3542
     -3542
         0
```

1) $506 \times \underline{8} = 4048$ Put 8 in quotient (q)
2) $451 - 448 = 3$
3) Bring down the 4
4) $506 \times \underline{0} = 0$ Put 0 in quotient (q)
5) $354 - 0 = 354$
6) Bring down the 2
7) $506 \times \underline{7} = 3542$ Put 7 in quotient (q)
8) $3542 - 3542 = 0$

Grade 6 Chapter 1 Lesson 10
Division Numbers 3

Review 1 A

Divide numbers

	a	b	c	d
# 1	1275 ÷ 25	1300 ÷ 25	1325 ÷ 25	1350 ÷ 25
# 2	1430 ÷ 26	1456 ÷ 26	1482 ÷ 26	1508 ÷ 26
# 3	1647 ÷ 27	1674 ÷ 27	1701 ÷ 27	1728 ÷ 27
# 4	1820 ÷ 28	1848 ÷ 28	1876 ÷ 28	1904 ÷ 28
# 5	2059 ÷ 29	2088 ÷ 29	2117 ÷ 29	2146 ÷ 29

Review 1 B

Divide numbers

	a	b	c	d
# 6	3162 ÷ 51	3213 ÷ 51	3264 ÷ 51	3315 ÷ 51
# 7	3366 ÷ 51	3417 ÷ 51	3468 ÷ 51	3519 ÷ 51
# 8	3224 ÷ 52	3276 ÷ 52	3328 ÷ 52	3380 ÷ 52
# 9	3432 ÷ 52	3484 ÷ 52	3536 ÷ 52	3588 ÷ 52
# 10	3286 ÷ 53	3339 ÷ 53	3392 ÷ 53	3445 ÷ 53

Grade 6 Chapter 1 Lesson 11
Order of Operations with Numbers 1

Addition and Subtraction Numbers

Addition of numbers

To add numbers, add in any order

Example 1 7000 + 35000 + 3000 = 7000 + 38000 = 45000

7000 + 35000 + 3000 = 42000 + 3000 = 45000

7000 + 35000 + 3000=7000 + 3000 + 35000=10000 + 35000=45000

$$\begin{array}{r} 35000 \\ +\ 7000 \\ \underline{3000} \\ 45000 \end{array}$$

Addition and Subtraction of numbers

To add and subtract numbers work from left to right

Example 2 7000 + 35000 - 3000 = 42000 - 3000 = 39000

1) $\begin{array}{r} 35000 \\ \underline{+\ 7000} \\ 42000 \end{array}$ 2) $\begin{array}{r} 42000 \\ \underline{-3000} \\ 39000 \end{array}$

Subtraction and Addition of numbers

To subtract and add numbers work from left to right

Example 3 35000 – 3000 + 7000 = 32000 + 7000 = 39000

1) $\begin{array}{r} 35000 \\ \underline{-\ 3000} \\ 32000 \end{array}$ 2) $\begin{array}{r} 32000 \\ \underline{+7000} \\ 39000 \end{array}$

Subtraction of numbers

To subtract numbers, work from left to right

Example 4 35000 – 3000 - 7000 = 32000 - 7000 = 25000

1) $\begin{array}{r} 35000 \\ \underline{-\ 3000} \\ 32000 \end{array}$ 2) $\begin{array}{r} 32000 \\ \underline{-\ 7000} \\ 25000 \end{array}$

Grade 6 Chapter 1 Lesson 11
Order of Operations with Numbers 1

Multiplication of Numbers

To multiply numbers, work in any order

$405 \times 8 \times 9 = 3240 \times 9 = 29160$ $405 \times 8 \times 9 = 405 \times 72 = 29160$

$506 \times 9 \times 8 = 4554 \times 8 = 36432$ $506 \times 9 \times 8 = 506 \times 72 = 36432$

Multiplication and Division of numbers

To multiply and divide numbers work from left to right

$309 \times 81 \div 9 = 25029 \div 9 = 2781$ $409 \times 81 \div 9 = 33129 \div 9 = 3681$

Division and Multiplication of numbers

To divide and multiply numbers work from left to right

$81 \div 9 \times 309 = 9 \times 309 = 2781$ $81 \div 9 \times 409 = 9 \times 409 = 3681$

Division of numbers

To divide numbers, work from left to right

$64000 \div 80 \div 8 = 800 \div 8 = 100$ $81000 \div 90 \div 9 = 900 \div 9 = 100$

Addition and Multiplication with Parentheses

Always first, work in parentheses

$(17 + 19) + 34 = 36 + 34 = 70$ $43 + (18 + 19) = 43 + 37 = 80$

$(25 + 5) \times 8 = 30 \times 8 = 240$ $6 \times (125+25) = 6 \times 150 = 900$

Addition and Subtraction with Parentheses

Always first, work in parentheses

$(700+300) -300=1000-300=700$ $(700-300) + 300 =400+300 =700$

$9000-(500+500) =9000- 1000=8000$ $(1200-500) -500=700-500=200$

Multiplication and Division with Parentheses

Always first, work in parentheses

$706 \times (36\div4) =706\times9=6354$ $24028 \div (16\div4) =24028\div4 =6007$

Expressions with Parentheses:

Always first, work in parentheses

$15 \times (807-200) =15\times607=9105$ $(9000+105) \div15 =9105\div15=607$

$(17472-9000) \div 12=8472\div12=706$ $84072\div(10+ 2)=84072\div12=7006$

Do multiplication or division before addition or subtraction

Grade 6 Chapter 1 Lesson 11

Order of Operations with Numbers 2

Multiplication and Addition of numbers

To multiply and add numbers always first multiply then add

Example 1 $5006 \times 5 + 4970 = 30000$

1) $5006 \times 5 = 25030$

$$\begin{array}{r} 5006 \\ \times \underline{\quad 5} \\ 25030 \end{array}$$

2) $25030 + 4970 = 30000$

$$\begin{array}{r} 25030 \\ \underline{+\ 4970} \\ 30000 \end{array}$$

Example 2 $1960 + 6005 \times 8 = 50000$

1)

$$\begin{array}{r} 6005 \\ \times \underline{\quad 8} \\ 48040 \end{array}$$

2)

$$\begin{array}{r} 48040 \\ \underline{+1960} \\ 50000 \end{array}$$

Multiplication and Subtraction of numbers

To multiply and subtract numbers always first multiply then subtract

Example 3 $506 \times 80 - 30480 = 10000$

1)

$$\begin{array}{r} 506 \\ \underline{\times\ 80} \\ 40480 \end{array}$$

2)

$$\begin{array}{r} 40480 \\ \underline{-\ 30480} \\ 10000 \end{array}$$

Example 4 $55600 - 608 \times 75 = 10000$

1)

$$\begin{array}{r} 608 \\ \underline{\times 75} \\ 3040 \\ \underline{42560} \\ 45600 \end{array}$$

2)

$$\begin{array}{r} 55600 \\ \underline{-\ 45600} \\ 10000 \end{array}$$

Grade 6 Chapter 1 Lesson 11

Order of Operations with Numbers 2

Division and Addition numbers

To divide and add numbers always first divide then add

Example 5

$$\overset{2}{7696+}\overset{1}{11520\div 5}=10000$$

2304

1) $5\overline{)11520}$ 2) 7696

```
 2304
5)11520        7696
 -10         + 2304
   15         10000
  -15
    20
   -20
```

Division and Subtraction numbers

To divide and subtract numbers, always first divide then subtract

Example 6

$$\overset{2}{907\ -}\ \overset{1}{45192\div 56}=100$$

```
    807
56)45192       907
  -448        -807
    392        100
   -392
```

Example 7

$6\div 2=3$ dividend 6 divisor 2 quotient 3

Every time bring down just one digit

$$907-307142\div 506=907-607=300$$

1) $506\times 6=3036$

2) $3071-3036=35$

3) $506\times 7=3542$

```
      607
506)307142
   -3036
      3542
     -3542
```

Grade 6 Chapter 1 Lesson 11
Order of Operations with Numbers 3
Multiplication Numbers
To multiply numbers, work in any order

Example 1 $405 \times 8 \times 9 = 3240 \times 9 = 29160$

1) 405
$\underline{\times 8}$
3240

2) 3240
$\underline{\times 9}$
29160

Example 2 $405 \times 8 \times 9 = 405 \times 72 = 29160$

1) 405
$\underline{\times 72}$
810
$\underline{28350}$
29160

Multiplication and Division Numbers
To multiply and divide numbers work from left to right

Example 3 $309 \times 81 \div 9 = 25029 \div 9 = 2781$

1) 309
$\underline{\times 81}$
309
$\underline{+24720}$
25029

2) $9\overline{)25029}$ = 3681
$\underline{-18}$
70
$\underline{-63}$
72
$\underline{-72}$
9
$\underline{-9}$

Example 4 $208 \times 24 \div 6 = 832$

1) 208
$\underline{\times 24}$
832
$\underline{+416}$
4992

2) $6\overline{)4992}$ = 832
$\underline{-48}$
19
$\underline{-18}$
12
$\underline{-12}$

Grade 6 Chapter 1 Lesson 11
Order of Operations with Numbers 3

Multiplication and Addition of numbers

To multiply and add numbers always first multiply then add

Example 5

$5006 \times 5 + 4970 = 30000$

1) $\begin{array}{r} 5006 \\ \times\ \underline{\ \ 5} \\ 25030 \end{array}$ 2) $\begin{array}{r} 25030 \\ \underline{+\ 4970} \\ 30000 \end{array}$

Example 6

$1960 + 6005 \times 8 = 50000$

1) $\begin{array}{r} 6005 \\ \times\ \underline{\ \ 8} \\ 48040 \end{array}$ 2) $\begin{array}{r} 48040 \\ \underline{+1960} \\ 50000 \end{array}$

Multiplication and Subtraction of numbers

To multiply and subtract numbers always first multiply then subtract

Example 7

$506 \times 80 - 30480 = 10000$

1) $\begin{array}{r} 506 \\ \underline{\times\ 80} \\ 40480 \end{array}$ 2) $\begin{array}{r} 40480 \\ \underline{-\ 30480} \\ 10000 \end{array}$

Example 8

$55600 - 608 \times 75 = 10000$

1) $\begin{array}{r} 608 \\ \underline{\times 75} \\ 3040 \\ +\underline{42560} \\ 45600 \end{array}$ 2) $\begin{array}{r} 55600 \\ \underline{-\ 45600} \\ 10000 \end{array}$

Grade 6 Chapter 1 Lesson 11
Order of Operations with Numbers 4
Division and Addition of numbers
To divide and add numbers always first divide then add

Example 1 $193230 \div 95 + 7966 = 10000$

1) 2034

95) 193230
-190
323
-285
380
-380

2) 7966 +2034=10000

Example 2 $7696 + 218880 \div 95 = 10000$

1) 2304

95)218880
-190
288
-285
380
-380

2) 7696 +2304=10000

Division and Subtraction of numbers
To divide and subtract numbers always first divide after then subtract

Example 3 $462060 \div 765 - 104 = 500$

1) 604

765) 462060
-4590
3060
-3060

2) 604-104=500

Example 4 $1307 - 462060 \div 765 = 500$

1) 807

706)569742
-5648
4942
-4942

2) 1307-807=500

Grade 6 Chapter 1 Lesson 11
Order of Operations with Numbers 4

Expressions with Parentheses
Always first, work in parentheses

Example 5 $15 \times (807 - 200) = 15 \times 607 = 9105$

1) $807 - 200 = 607$

```
 807
-200
----
 607
```

2)

```
  607
  ×15
 ----
 3035
+6070
-----
 9105
```

Example 6 $(200 + 105) \times 15 = 305 \times 15 = 4575$

1)

```
 200
+105
----
 305
```

2)

```
  305
  ×15
 ----
 1525
 3050
 ----
 4575
```

Example 7 $(17472 - 9000) \div 12 = 8472 \div 12 = 706$

1) 17472-9000=8472

2)

```
     706
12)8472
   -84
   ---
     72
    -72
    ---
```

Example 8 $84072 \div (10 + 2) = 84072 \div 12 = 7006$

1) $10 + 2 = 12$

2)

```
    7006
12)84072
   -84
   ---
     072
    -072
    ----
```

Grade 6 Chapter 1 Lesson 11
Order of Operations with Numbers 5

1) Multiplication numbers: for memory <u>(x, x, any)</u>
To multiply numbers, work in any order
$9 \times 8 \times 7 = 72 \times 7 = 504$ $\quad$ $9 \times 8 \times 7 = 9 \times 56 = 504$

2) Work from left to right: for memory <u>(L → R)</u>

a) Multiplication and Division numbers: (× or ÷)
To multiply and divide numbers work from left to right (L → R)
$9 \times 6 \div 3 = 54 \div 3 = 18$ $\quad$ $4 \times 81 \div 9 = 324 \div 9 = 36$

b) Division and Multiplication numbers: (÷ or ×)
To divide and multiply numbers work from left to right <u>(L → R)</u>
$24 \div 6 \times 2 = 4 \times 2 = 8$ $\quad$ $81 \div 9 \times 4 = 9 \times 4 = 36$

c) Division numbers: (÷, ÷)
To divide numbers, work from left to right <u>(L → R)</u>
$24 \div 4 \div 2 = 6 \div 2 = 3$ $\quad$ $810 \div 9 \div 3 = 90 \div 9 = 10$

3) Do Multiplication or Division before Addition or Subtraction
for memory (× or ÷) before (+ or -)
a) Multiplication and Addition numbers: 1) × 2) +
To multiply and add numbers always first multiply than add
$9 \times 6 + 9 = 54 + 9 = 63$ $\quad$ $9 + 9 \times 9 = 9 + 81 = 90$

b) Multiplication and Subtraction numbers: 1) × 2) −
To multiply and subtract numbers always first multiply than subtract
$9 \times 9 - 9 = 81 - 9 = 72$ $\quad$ $90 - 9 \times 9 = 90 - 81 = 9$

c) Division and Addition numbers: 1) ÷ 2) +
To divide and add numbers always first divide than add
$50 \div 5 + 39 = 10 + 39 = 49$ $\quad$ $88 + 171 \div 9 = 88 + 19 = 107$

d) Division and Subtraction numbers: 1) ÷ 2) −
To divide and subtract numbers always first divide than subtract
$108 \div 6 - 9 = 18 - 9 = 9$ $\quad$ $500 - 126 \div 7 = 500 - 18 = 482$

Grade 6 Chapter 1 Lesson 11
Order of Operations with Numbers 5

Review 1 A
Perform the operations

	a	b
# 1	$98 \times 4 \times 5$	$96 \times 8 \times 5$
# 2	$96 \times 40 \div 8$	$98 \times 45 \div 9$
# 3	$35 \div 7 \times 986$	$30 \div 6 \times 978$
# 4	$7848 \div 9 \div 8$	$6048 \div 8 \div 7$
# 5	$985 \times 8 + 120$	$120 + 976 \times 5$

Review 1 B
Perform the operations

	a	b
# 6	$95 \times 98 - 10$	$96 \times 95 - 20$
# 7	$8200 - 85 \times 96$	$8000 - 84 \times 95$
# 8	$7264 \div 8 + 91$	$5436 \div 6 + 93$
# 9	$8544 \div 96 - 19$	$8526 \div 98 - 27$
# 10	$9500 - 9310 \div 95$	$9310 - 9120 \div 95$

Grade 6 Chapter 1 Lesson 12
Signs of operations

Put + or – between numbers that each problem was correct

20 ? 60 = 80 20 + 60 = 80 60 ? 20 = 40 60 - 20 = 40

20 ? 70 = 90 20 + 70 = 90 90 ? 20 = 70 90 - 20 = 70

Review 1 A

	a	b	c	d
# 1	20 ? 40=60	30 ? 30=60	40 ? 20=60	50 ? 10=60
# 2	60 ? 10=50	70 ? 10=60	80 ? 10=70	90 ? 10=80
# 3	21 ? 42=63	31 ? 32=63	41 ? 22=63	51 ? 12=63
# 4	61 ? 12=49	71 ? 12=59	81 ? 12=69	91 ? 12=79
# 5	22 ? 44=66	32 ? 34=66	42 ? 24=66	52 ? 14=66
# 6	62 ? 14=48	72 ? 14=58	82 ? 14=68	92 ? 14=78
# 7	23 ? 46=69	33 ? 36=69	43 ? 26=69	53 ? 16=69
# 8	63 ? 16=47	73 ? 16=57	83 ? 16=67	93 ? 16=77
# 9	24 ? 48=72	34 ? 38=72	44 ? 28=72	54 ? 18=72
# 10	64 ? 18=46	74 ? 18=56	84 ? 18=66	94 ? 18=76
# 11	25 ? 41=66	35 ? 31=66	45 ? 21=66	55 ? 11=66
# 12	65 ? 11=54	75 ? 11=64	85 ? 11=74	95 ? 11=84
# 13	26 ? 43=69	36 ? 33=69	46 ? 23=69	56 ? 13=69
# 14	66 ? 13=53	76 ? 13=63	86 ? 13=73	96 ? 13=83
# 15	27 ? 45=72	37 ? 35=72	47 ? 25=72	57 ? 15=72
# 16	67 ? 15=52	77 ? 15=62	87 ? 15=72	97 ? 15=82
# 17	28 ? 47=75	38 ? 37=75	48 ? 27=75	58 ? 17=75
# 18	68 ? 17=51	78 ? 17=61	88 ? 17=71	98 ? 17=81
# 19	29 ? 49=78	39 ? 39=78	49 ? 29=78	59 ? 19=78
# 20	69 ? 19=50	79 ? 19=60	89 ? 19=70	99 ? 19=80

Grade 6 Chapter 1 Lesson 12

Signs of operations

Put × or ÷ between numbers that each problem was correct

6 ? 2 = 12	6 × 2 = 12	6 ? 2 = 3	6 ÷ 2 = 3
9 ? 3 = 27	9 × 3 = 27	9 ? 3 = 3	9 ÷ 3 = 3

Review 1 B

	a	b	c	d
# 1	6 ? 0=0	7 ? 0=0	8 ? 0=0	9 ? 0=0
# 2	6 ? 6=1	7 ? 7=1	8 ? 8=1	9 ? 9=1
# 3	6 ? 2=12	7 ? 2=14	8 ? 2=16	9 ? 2=18
# 4	12 ? 6=2	14 ? 7=2	16 ? 8=2	18 ? 9=2
# 5	6 ? 3=18	7 ? 3=21	8 ? 3=24	9 ? 3=27
# 6	18 ? 6=3	21 ? 7=3	24 ? 8=3	27 ? 9=3
# 7	6 ? 4=24	7 ? 4=28	8 ? 4=32	9 ? 4=36
# 8	24 ? 6=4	28 ? 7=4	32 ? 8=4	36 ? 9=4
# 9	6 ? 5=30	7 ? 5=35	8 ? 5=40	9 ? 5=45
# 10	30 ? 6=5	35 ? 7=5	40 ? 8=5	45 ? 9=5
# 11	6 ? 6=36	7 ? 6=42	8 ? 6=48	9 ? 6=54
# 12	36 ? 6=6	42 ? 7=6	48 ? 8=6	54 ? 9=6
# 13	6 ? 7=42	7 ? 7=49	8 ? 7=56	9 ? 7=63
# 14	42 ? 6=7	49 ? 7=7	56 ? 8=7	63 ? 9=7
# 15	6 ? 8=48	7 ? 8=56	8 ? 8=6 4	9 ? 8=72
# 16	48 ? 6=8	56 ? 7=8	64 ? 8=8	72 ? 9=8
# 17	6 ? 9=54	7 ? 9=63	8 ? 9=72	9 ? 9=81
# 18	54 ? 6=9	63 ? 7=9	72 ? 8=9	81 ? 9=9
# 19	6 ? 10=60	7 ? 10=70	8 ? 10=80	9 ? 10=90
# 20	60 ? 6=10	70 ? 7=10	80 ? 8=10	90 ? 9=10

Grade 6 Chapter 1 Lesson 13
Reviews for Test 1
Review 1 A

Each problem is worth 5 points **Show all steps for full credits**

Part 1

Evaluate

1) 900000
   - 456789

1)______

2) 987
   ×806

2)______

3) 908
   ×605

3)______

4) 437658 ÷ 543

4)______

5) 304920 ÷ 504

5)______

Grade 6 Chapter 1 Lesson 13

Review 1 A

Each problem is worth 5 points **Show all steps for full credits**

Part 2

Evaluate

6) $300300 \div 60 \times 5$

6)______

7) $9250 - 650 \times 5$

7)______

8) $90081 - 81072 \div 9$

8)_______

9) $(50 \times 300) \div 6$

9)______

10) $361 \div (117 - 98)$

10)______

Grade 6 Chapter 1 Lesson 13

Review 1 B

Each problem is worth 5 points **Show all steps for full credits**

Part 1

Calculate

1) $\begin{array}{r} 800000 \\ -\ 345678 \\ \hline \end{array}$

1)______

2) $\begin{array}{r} 876 \\ \times 705 \\ \hline \end{array}$

2)______

3) $\begin{array}{r} 807 \\ \times 706 \\ \hline \end{array}$

3)______

4) $304560 \div 432$

4)______

5) $284518 \div 403$

5)______

Grade 6 Chapter 1 Lesson 13

Review 1 B

Each problem is worth 5 points **Show all steps for full credits**

Part 2

Evaluate

6) $280280 \div 70 \times 4$

6)______

7) $8960 - 740 \times 4$

7)______

8) $72064 - 64056 \div 8$

8)______

9) $(40 \times 420) \div 7$

9)______

10) $324 \div (105 - 87)$

10)______

Grade 6 Chapter 1 Lesson 14

Test 1 A

Each problem is worth 5 points **Show all steps for full credits**

Part 1

Evaluate

1) $\begin{array}{r} 700000 \\ -\ 234567 \\ \hline \end{array}$

1)______

2) $\begin{array}{r} 765 \\ \times 604 \\ \hline \end{array}$

2)______

3) $\begin{array}{r} 706 \\ \times 605 \\ \hline \end{array}$

3)______

4) $193884 \div 321$

4)______

5) $243714 \div 302$

5)______

Grade 6 Chapter 1 Lesson 14

Test 1 A

Each problem is worth 5 points **Show all steps for full credits**

Part 2

Evaluate

6) $240240 \div 80 \times 3$

6)______

7) $7490 - 830 \times 3$

7)______

8) $56049 - 49042 \div 7$

8)______

9) $(30 \times 560) \div 8$

9)______

10) $289 \div (93 - 76)$

10)______

Grade 6 Chapter 1 Lesson 14

Test 1 B

Each problem is worth 5 points **Show all steps for full credits**

Part 1

Evaluate

1) $\begin{array}{r} 600000 \\ -\ 123456 \\ \hline \end{array}$

1)______

2) $\begin{array}{r} 654 \\ \times 503 \\ \hline \end{array}$

2)______

3) $\begin{array}{r} 605 \\ \times 403 \\ \hline \end{array}$

3)_______

4) $105630 \div 210$

4)_______

5) $182508 \div 201$

5)_______

Grade 6 Chapter 1 Lesson 14

Test 1 B

Each problem is worth 5 points **Show all steps for full credits**

Part 2

Evaluate

6) $180180 \div 90 \times 2$

6)______

7) $6840 - 920 \times 2$

7)______

8) $42036 - 36030 \div 6$

8)______

9) $(20 \times 720) \div 9$

9)______

10) $256 \div (81 - 65)$

10)______

Grade 6 Answers Chapter 1

Lesson 1 Four – digit numbers

Review 1 A

	a	b	c	d	e
# 1	3110	4110	5110	6110	7554
# 2	4221	5221	6221	7221	8554
# 3	5332	6332	7332	8665	9776
# 4	6443	7443	8443	9665	9887

Review 1 B	a	b	c	d	e
# 5	1191	1282	1373	1464	1595
# 6	1292	1383	1474	1565	1686
# 7	1393	1484	1575	1696	1797
# 8	1494	1585	1676	1787	898

Lesson 2 Five – digit numbers

Review 1 A

	a	b	c	d	e
# 1	31100	41100	51100	61100	75540
# 2	42210	52210	62210	72210	85540
# 3	53320	63320	73320	86650	97760
# 4	64430	74430	84430	96650	98870

Review 1 B	a	b	c	d	e
# 5	11910	12820	13730	14640	15950
# 6	12920	13830	14740	15650	16860
# 7	13930	14840	15750	16960	17970
# 8	14940	15850	16760	17870	8980

Lesson 3 Six-digit numbers

Review 1 A

	a	b	c	d	e
1)	311000	411000	511000	611000	755400
2)	422100	522100	622100	722100	855400
3)	533200	633200	733200	866500	977600
4)	644300	744300	844300	966500	988700

Review 1 B

5)	119100	128200	137300	146400	159500
6)	129200	138300	147400	156500	168600
7)	139300	148400	157500	169600	179700
8)	149400	158500	167600	178700	89800

Grade 6 Answers Chapter 1

Lesson 4 Seven-digit numbers Review 1 A

	a	b	c	d
1)	3,110,000	4,110,000	5,110,000	6,110,000
2)	4,221,000	5,221,000	6,221,000	7,221,000
3)	5,332,000	6,332,000	7,332,000	8,665,000
4)	6,443,000	7,443,000	8,443,000	9,665,000
5)	7,554,000	8,554,000	9,776,000	9,887,000

Review 1 B

6)	1,191,000	1,282,000	1,373,000	1,464,000
7)	1,292,000	1,383,000	1,474,000	1,565,000
8)	1,393,000	1,484,000	1,575,000	1,696,000
9)	1,494,000	1,585,000	1,676,000	1,787,000
10)	1,595,000	1,686,000	1,797,000	898,000

Lesson 6 Multiply Numbers. Review 1 A

	a	b	c	d	e
# 1	2750	2805	2860	2915	2970
# 2	3025	3080	3135	3190	3245
# 3	2800	2856	2912	2968	3024
# 4	3080	3136	3192	3248	3304
# 5	2850	2907	2964	3021	3078

Review 1 B	a	b	c	d	e
# 6	3300	3355	3410	3465	3520
# 7	3575	3630	3685	3740	3795
# 8	3360	3416	3472	3528	3584
# 9	3640	3696	3752	3808	3864
# 10	3420	3477	3534	3591	3648

Lesson 6 Multiply numbers. Review 2 A

	a	b	c	d
# 1	2670	2835	3936	4068
# 2	5215	5523	6848	6936
# 3	8703	8802	2715	2880
# 4	4050	4122	5278	5586
# 5	6920	7008	8784	8865

Review 2 B	a	b	c	d
# 1	2730	3290	3870	4134
# 2	5292	5376	6936	7032
# 3	8802	8883	2820	3425
# 4	3924	4188	5348	5502
# 5	7008	7176	8703	8784

Grade 6 Answers Chapter 1

Lesson 9 Review 1 A

	a	b
# 1	546	534
# 2	578	567
# 3	645	656
# 4	689	678
# 5	756	745
# 6	769	789
# 7	867	834
# 8	878	856
# 9	978	967
# 10	987	986

Review 1 B

	a	b
# 1	564	543
# 2	587	576
# 3	654	675
# 4	698	687
# 5	765	754
# 6	796	798
# 7	875	846
# 8	896	865
# 9	967	976
# 10	978	989

Lesson 10 Review 1 A

	a	b	c	d
# 1	51	52	53	54
# 2	55	56	57	58
# 3	61	62	63	64
# 4	65	66	67	68
# 5	71	72	73	74

Review 1 B

	a	b	c	d
# 6	62	63	64	65
# 7	66	67	68	69
# 8	62	63	64	65
# 9	66	67	68	69
# 10	62	63	64	65

Lesson 11 Order of Operations 5

Review 1 A

	a	b
# 1	1960	3840
# 2	480	490
# 3	4930	4890
# 4	109	108
# 5	8000	5000

Review 1 B

	a	b
# 6	9300	9100
# 7	40	20
# 8	999	999
# 9	70	60
# 10	9402	9214

Lesson 12 Signs of operations. Review 1 A

	a	b	c	d
# 1	+	+	+	+
# 2	-	-	-	-
# 3	+	+	+	+
# 4	-	-	-	-
# 5	+	+	+	+
# 6	-	-	-	-
# 7	+	+	+	+
# 8	-	-	-	-
# 9	+	+	+	+
#10	-	-	-	-

	a	b	c	d
# 11	+	+	+	+
# 12	-	-	-	-
# 13	+	+	+	+
# 14	-	-	-	-
# 15	+	+	+	+
# 16	-	-	-	-
# 17	+	+	+	+
# 18	-	-	-	-
# 19	+	+	+	+
# 20	-	-	-	-

Grade 6 Answers Chapter 1

Lesson 12 Review 1 B

	a	b	c	d		a	b	c	d
# 1	×	×	×	×	#11	×	×	×	×
#2	÷	÷	÷	÷	#12	÷	÷	÷	÷
#3	×	×	×	×	#13	×	×	×	×
#4	÷	÷	÷	÷	#14	÷	÷	÷	÷
#5	×	×	×	×	#15	×	×	×	×
#6	÷	÷	÷	÷	#16	÷	÷	÷	÷
#7	×	×	×	×	#17	×	×	×	×
#8	÷	÷	÷	÷	#18	÷	÷	÷	÷
#9	×	×	×	×	#19	×	×	×	×
#10	÷	÷	÷	÷	#20	÷	÷	÷	÷

Lesson 13 Review 1 A

1)	443211	2)	795522	3)	549340	4)	806	5)	605
6)	25025	7)	6000	8)	81073	9)	2500	10)	19

Review 1 B

1)	454322	2)	617580	3)	569742	4)	705	5)	706
6)	16016	7)	6000	8)	64057	9)	2400	10)	18

Lesson 14 Test 1 A

1)	465433	2)	462060	3)	427130	4)	604	5)	807
6)	9009	7)	5000	8)	49043	9)	2100	10)	17

Test 1 B

1)	476544	2)	328962	3)	243815	4)	503	5)	908
6)	4004	7)	5000	8)	36031	9)	1600	10)	16

Grade 6

Chapter 2

Multiplication

Numbers 3

Grade 6 Chapter 2 Contents
Multiplication Numbers
Lesson 1 – 10
Basic Problems

Work with lesson 1, read rules and examples, 2) Solve the problems from contents, 3) Check answers in the end of chapter 1, 4) Work with the next lessons 2, 3, 4, 5, …, 5) Solve Test 1 A, 6) Grading Scale: A =10 points, B=9points, C=8 points, D=7 points, F=0-6 points, 7) If you solve 10 problems correctly, you get an A, 8) If have an A, you can move on to next chapter, 9) If your grade is lower than an A, work with the mistakes, then solve Test 1B, 10) When you finish all of the chapters, solve Final Exam in last chapter.

Grade 6 Chapter 2 Lesson 1
Multiplication Numbers 1

Third operation with numbers is multiplication × or ·
Multiplication is the operation of repeated addition
$2\times 3 = 2 + 2 + 2 = 6$ or $2 \cdot 3 = 2 + 2 + 2 = 6$

First number in multiplication called the factor 2
Second number in multiplication called the factor 3
Result of multiplication called the product 6
Factor × Factor = Product $2 \times 3 = 6$

Properties of multiplication

Two numbers can multiply in any order
The product will be the same AB = BA
$2 \times 3 = 2 + 2 + 2 = 6$ $3 \times 2 = 3 + 3 = 6$
The product of a number and zero is zero
$A \times 0 = 0 \times A = 0$ $5 \times 0 = 0$ $0 \times 5 = 0$
The product of a number and one is the same number
$A \times 1 = 1 \times A = A$ $5 \times 1 = 5$ $1 \times 5 = 5$

Memorize the Table of Multiplication
Number 5 from the left column multiply by number 4 from first row to get answer 20

×	2	3	<u>4</u>	5	6	7	8	9
1	2	3	4	5	6	7	8	9
2	4	6	8	10	12	14	16	18
3	6	9	12	15	18	21	24	27
4	8	12	16	20	24	28	32	36
<u>5</u>	10	15	<u>20</u>	25	30	35	40	45
6	12	18	24	30	36	42	48	54
7	14	21	28	35	42	49	56	63
8	16	24	32	40	48	56	64	72
9	18	27	36	45	54	63	72	81
10	20	30	40	50	60	70	80	90

Grade 6 Chapter 2 Lesson 1
Multiplication Numbers 1
Factor × Factor = Product $2 \times 3 = 6$
Example

1) $15\times 9 = (10 + 5) \times 9 =$
2) $15 = 10 + 5$
3) $10 \times 9 = 90$ (Product 1 = P 1)
4) $5 \times 9 = 45$ (Product 2 = P 2)
5) $90 + 45 = 135$ (Product = P)
6) $15\times 9 = 135$

Review 1 A
Multiply numbers in row

	a	b	c	d	e
1)	10×2	11×2	12×2	13×2	14×2 = (10 + 4) ×2
2)	10×3	11×3	12×3	13×3	14×3
3)	10×4	11×4	12×4	13×4	14×4
4)	10×5	11×5	12×5	13×5	14×5
5)	10×6	11×6	12×6	13×6	14×6
6)	10×7	11×7	12×7	13×7	14×7
7)	10×8	11×8	12×8	13×8	14×8
8)	10×9	11×9	12×9	13×9	14×9

Review 1 B
Multiply numbers in column

	a	b	c	d	e
1)	15×2	16×2	17×2	18×2	19×2= (10+9)×2
2)	15×3	16×3	17×3	18×3	19×3
3)	15×4	16×4	17×4	18×4	19×4
4)	15×5	16×5	17×5	18×5	19×5
5)	15×6	16×6	17×6	18×6	19×6
6)	15×7	16×7	17×7	18×7	19×7
7)	15×8	16×8	17×8	18×8	19×8
8)	15×9	16×9	17×9	18×9	19×9

Grade 6 Chapter 2 Lesson 1
Multiplication Numbers 1
Example

1056	1)	1056 = 1000 + 50 + 6
× 5	2)	5× 6 = 30
30	3)	5×50 = 250
250	4)	5× 1000 = 5000
+ 5000		
5280		

Review 2 A
Multiply

	a	b	c	d	e
# 1	1020 × 2	1022 × 2	1024 × 2	1026 × 2	1028 × 2
# 2	1031 × 3	1033 × 3	1035 × 3	1037 × 3	1039 × 3
# 3	1040 × 4	1042 × 4	1044 × 4	1046 × 4	1048 × 4
# 4	1051 × 5	1053 × 5	1055 × 5	1057 × 5	1059 × 5
# 5	1600 × 6	1602 × 6	1604 × 6	1606 × 6	1608 × 6
# 6	1701 × 7	1703 × 7	1705 × 7	1707 × 7	1709 × 7
# 7	1800 × 8	1802 × 8	1804 × 8	1806 × 8	1808 × 8
# 8	1901 × 9	1903 × 9	1905 × 9	1907 × 9	1909 × 9

Grade 6 Chapter 2 Lesson 1
Multiplication Numbers 1
Example

$$\begin{array}{r} 2056 \\ \times\ \ 5 \\ \hline 30 \\ 250 \\ +\ 10000 \\ \hline 10280 \end{array}$$

1) 2056 = 2000 + 50 + 6
2) 5×6 = 30
3) 5×50 = 250
4) 5× 2000 = 10000

Review 2 B
Multiply

	a	b	c	d	e
# 1	2020 × 2	2022 × 2	2024 × 2	2026 × 2	2028 × 2
# 2	2031 × 3	2033 × 3	2035 × 3	2037 × 3	2039 × 3
# 3	2040 × 4	2042 × 4	2044 × 4	2046 × 4	2048 × 4
# 4	2051 × 5	2053 × 5	2055 × 5	2057 × 5	2059 × 5
# 5	2600 × 6	2602 × 6	2604 × 6	2606 × 6	2608 × 6
# 6	2701 × 7	2703 × 7	2705 × 7	2707 × 7	2709 × 7
# 7	2800 × 8	2802 × 8	2804 × 8	2806 × 8	2808 × 8
# 8	2901 × 9	2903 × 9	2905 × 9	2907 × 9	2909 × 9

Grade 6 Chapter 2 Lesson 2
Multiplication numbers 2
Example

$$\begin{array}{r} 3506 \\ \times\ \ 5 \\ \hline 30 \\ 2500 \\ +\ 15000 \\ \hline 17530 \end{array}$$

1) 3506 = 3000 + 500 + 6
2) 5×6 = 30
3) 5×500 = 2500
4) 5× 3000 = 15000

Review 1 A
Multiply

	a	b	c	d	e
# 1	3020 × 2	3022 × 2	3024 × 2	3026 × 2	3028 × 2
# 2	3031 × 3	3033 × 3	3035 × 3	3037 × 3	3039 × 3
# 3	3040 × 4	3042 × 4	3044 × 4	3046 × 4	3048 × 4
# 4	3051 × 5	3053 × 5	3055 × 5	3057 × 5	3059 × 5
# 5	3600 × 6	3602 × 6	3604 × 6	3606 × 6	3608 × 6
# 6	3701 × 7	3703 × 7	3705 × 7	3707 × 7	3709 × 7
# 7	3800 × 8	3802 × 8	3804 × 8	3806 × 8	3808 × 8
# 8	3901 × 9	3903 × 9	3905 × 9	3907 × 9	3909 × 9

Grade 6 Chapter 2 Lesson 2
Multiplication of numbers 2
Example

4506	1) 4506 = 4000 + 500 + 6
× 5	2) 5×6 = 30
30	3) 5×500 = 2500
2500	4) 5× 4000 = 20000
+ 20000	
22530	

Review 1 B
Multiply Numbers

	a	b	c	d	e
# 1	4020 × 2	4022 × 2	4024 × 2	4026 × 2	4028 × 2
# 2	4031 × 3	4033 × 3	4035 × 3	4037 × 3	4039 × 3
# 3	4040 × 4	4042 × 4	4044 × 4	4046 × 4	4048 × 4
# 4	4051 × 5	4053 × 5	4055 × 5	4057 × 5	4059 × 5
# 5	4600 × 6	4602 × 6	4604 × 6	4606 × 6	4608 × 6
# 6	4701 × 7	4703 × 7	4705 × 7	4707 × 7	4709 × 7
# 7	4800 × 8	4802 × 8	4804 × 8	4806 × 8	4808 × 8
# 8	4901 × 9	4903 × 9	4905 × 9	4907 × 9	4909 × 9

Grade 6 Chapter 2 Lesson 2
Multiplication Number 2
Example

5506	1)	5506 = 5000 + 500 + 6
× 5	2)	5×6 = 30
30	3)	5×500 = 2500
2500	4)	5× 5000 = 25000
+ 25000		
27530		

Review 2 A
Multiply

	a	b	c	d	e
# 1	5020 × 2	5022 × 2	5024 × 2	5026 × 2	5028 × 2
# 2	5031 × 3	5033 × 3	5035 × 3	5037 × 3	5039 × 3
# 3	5040 × 4	5042 × 4	5044 × 4	5046 × 4	5048 × 4
# 4	5051 × 5	5053 × 5	5055 × 5	5057 × 5	5059 × 5
# 5	5600 × 6	5602 × 6	5604 × 6	5606 × 6	5608 × 6
# 6	5701 × 7	5703 × 7	5705 × 7	5707 × 7	5709 × 7
# 7	5800 × 8	5802 × 8	5804 × 8	5806 × 8	5808 × 8
# 8	5901 × 9	5903 × 9	5905 × 9	5907 × 9	5909 × 9

Grade 6 Chapter 2 Lesson 2
Multiplication Numbers 2

Factor × Factor = Product (P)   2×3 = 6 (P)

Example

Top 3 2	1) 5 × 4 = 2<u>0</u>	7) 5 × 7 = 3<u>5</u>
6704	2) Put <u>0</u> below 5 (P)	8) Put <u>5</u> ahead 2 (P)
× 5	3) Put 2 above 0 (top)	9) Put 3 above 6 (top)
33520 (P)	4) 5 × 0 = 0	10) 5 × 6 = 30
	5) 2 + 0 = <u>2</u>	11) 30 + 3 = <u>33</u>
	6) Put <u>2</u> ahead 0 (P)	12) Put <u>33</u> ahead 5 (P)

Review 2 B

Multiply

	a	b	c	d	e
# 1	6020 × 2	6022 × 2	6024 × 2	6026 × 2	6028 × 2
# 2	6031 × 3	6033 × 3	6035 × 3	6037 × 3	6039 × 3
# 3	6040 × 4	6042 × 4	6044 × 4	6046 × 4	6048 × 4
# 4	6051 × 5	6053 × 5	6055 × 5	6057 × 5	6059 × 5
# 5	6600 × 6	6602 × 6	6604 × 6	6606 × 6	6608 × 6
# 6	6701 × 7	6703 × 7	6705 × 7	6707 × 7	6709 × 7
# 7	6800 × 8	6802 × 8	6804 × 8	6806 × 8	6808 × 8
# 8	6901 × 9	6903 × 9	6905 × 9	6907 × 9	6909 × 9

Grade 6 Chapter 2 Lesson 3

Multiplication Numbers 3

Factor × Factor = Product (P)   2×3 = 6 (P)

Example

```
Top  2 3      1) 5 × 6 = 30               7)  5 × 4 = 20
   7406       2) Put 0 below 5 (P)        8)  Put 0 ahead 3 (P)
 ×    5       3) Put 3 above 0 (top)      9)  Put 2 above 7 (top)
 37030 (P)    4) 5 × 0 = 0               10)  5 × 7 = 35
              5) 3 + 0 = 3               11)  35 + 2 = 37
              6) Put 3 ahead 0 (P)       12)  Put 37 ahead 0 (P)
```

Review 1 A

Multiply Numbers

	a	b	c	d	e
# 1	7020 × 2	7022 × 2	7024 × 2	7026 × 2	7028 × 2
# 2	7031 × 3	7033 × 3	7035 × 3	7037 × 3	7039 × 3
# 3	7040 × 4	7042 × 4	7044 × 4	7046 × 4	7048 × 4
# 4	7051 × 5	7053 × 5	7055 × 5	7057 × 5	7059 × 5
# 5	7600 × 6	7602 × 6	7604 × 6	7606 × 6	7608 × 6
# 6	7701 × 7	7703 × 7	7705 × 7	7707 × 7	7709 × 7
# 7	7800 × 8	7802 × 8	7804 × 8	7806 × 8	7808 × 8
# 8	7901 × 9	7903 × 9	7905 × 9	7907 × 9	7909 × 9

Grade 6 Chapter 2 Lesson 3
Multiplication Numbers 3

Factor × Factor = Product (P) **2×3 = 6 (P)**

Example

Top 2 3	1) 5×6=3<u>0</u>	7) 5×4=2<u>0</u>
8406	2) Put <u>0</u> below 5 (P)	8) Put <u>0</u> in ahead 3 (P)
× 5	3) Put 3 above 0 (top)	9) Put 2 above 8 (top)
42030 (P)	4) 5×0=0	10) 5×8=40
	5) 3+0=<u>3</u>	11) 40+2=<u>42</u>
	6) Put <u>3</u> ahead 0 (P)	12) Put <u>42</u> ahead 0 (P)

Review 1 B

Multiply

	a	b	c	d	e
# 1	8020 × 2	8022 × 2	8024 × 2	8026 × 2	8028 × 2
# 2	8031 × 3	8033 × 3	8035 × 3	8037 × 3	8039 × 3
# 3	8040 × 4	8042 × 4	8044 × 4	8046 × 4	8048 × 4
# 4	8051 × 5	8053 × 5	8055 × 5	8057 × 5	8059 × 5
# 5	8600 × 6	8602 × 6	8604 × 6	8606 × 6	8608 × 6
# 6	8701 × 7	8703 × 7	8705 × 7	8707 × 7	8709 × 7
# 7	8800 × 8	8802 × 8	8804 × 8	8806 × 8	8808 × 8
# 8	8901 × 9	8903 × 9	8905 × 9	8907 × 9	8909 × 9

Grade 6 Chapter 2 Lesson 3

Multiplication Numbers 3

Factor × Factor = Product (P)　　　2×3 = 6 (P)

Example

Top 2 3	1) 5 × 6 = 30	7) 5 × 4 = 20
9406	2) Put 0 below 5 (P)	8) Put 0 ahead 3 (P)
× 5	3) Put 3 above 0 (top)	9) Put 2 above 9 (top)
47030 (P)	4) 5 × 0 = 0	10) 5 × 9 = 45
	5) 3 + 0 = 3	11) 45 + 2 = 47
	6) Put 3 ahead 0 (P)	12) Put 47 ahead 0 (P)

Review 2 A

Multiply

	a	b	c	d	e
# 1	9020 × 2	9022 × 2	9024 × 2	9026 × 2	9028 × 2
# 2	9031 × 3	9033 × 3	9035 × 3	9037 × 3	9039 × 3
# 3	9040 × 4	9042 × 4	9044 × 4	9046 × 4	9048 × 4
# 4	9051 × 5	9053 × 5	9055 × 5	9057 × 5	9059 × 5
# 5	9600 × 6	9602 × 6	9604 × 6	9606 × 6	9608 × 6
# 6	9701 × 7	9703 × 7	9705 × 7	9707 × 7	9709 × 7
# 7	9800 × 8	9802 × 8	9804 × 8	9806 × 8	9808 × 8
# 8	9901 × 9	9903 × 9	9905 × 9	9907 × 9	9909 × 9

Grade 6 Chapter 2 Lesson 3
Multiplication Numbers 3
Example

```
  104    1) 15 = 10 + 5
× 15
  520    2) 104       3) 104 ×10 =1040
+1040       ×5
 1560       520
```

Review 2 B
Multiply

	a	b	c	d	e
# 1	100 ×12	101 ×12	102 ×12	103 ×12	104 × 12
# 2	105 ×13	106 ×13	107 ×13	108 ×13	109 ×13
# 3	100 ×14	101 ×14	102 ×14	103 ×14	104 ×14
# 4	105 ×15	106 ×15	107 ×15	108 ×15	109 ×15
# 5	100 × 16	101 × 16	102 × 16	103 × 16	104 × 16
# 6	105 ×17	106 ×17	107 ×17	108 ×17	109 ×17
# 7	100 ×18	101 ×18	102 ×18	103 ×18	104 ×18
# 8	105 ×19	106 ×19	107 ×19	108 ×19	109 ×19

Grade 6 Chapter 2 Lesson 4
Multiplication Numbers 4
Example

```
  204     1)  15 = 10 + 5
 × 15
 1020     2)  204      3)  204 ×10 =2040
+2040          ×5
 3060        1020
```

Review 1 A
Multiply

	a	b	c	d	e
# 1	200 ×12	201 ×12	202 ×12	203 ×12	204 × 12
# 2	205 ×13	206 ×13	207 ×13	208 ×13	209 ×13
# 3	200 ×14	201 ×14	202 ×14	203 ×14	204 ×14
# 4	205 ×15	206 ×15	207 ×15	208 ×15	209 ×15
# 5	200 × 16	201 × 16	202 × 16	203 × 16	204 × 16
# 6	205 ×17	206 ×17	207 ×17	208 ×17	209 ×17
# 7	200 ×18	201 ×18	202 ×18	203 ×18	204 ×18
# 8	205 ×19	206 ×19	207 ×19	208 ×19	209 ×19

Grade 6 Chapter 2 Lesson 4
Multiplication Numbers 4
Example

$$\begin{array}{r} 304 \\ \times\ 25 \\ \hline 1520 \\ +\ 6080 \\ \hline 7600 \end{array}$$

1) $25 = 20 + 5$

2) $\begin{array}{r} 304 \\ \times 5 \\ \hline 1520 \end{array}$

3) $\begin{array}{r} 304 \\ \times 20 \\ \hline 6080 \end{array}$

Review 1 B
Multiply

	a	b	c	d	e
# 1	300 ×22	301 ×22	302 ×22	303 ×22	304 × 22
# 2	305 ×23	306 ×23	307 ×23	308 ×23	309 ×23
# 3	300 ×24	301 ×24	302 ×24	303 ×24	304 ×24
# 4	305 ×25	306 ×25	307 ×25	308 ×25	309 ×25
# 5	300 × 26	301 × 26	302 × 26	303 × 26	304 × 26
# 6	305 ×27	306 ×27	307 ×27	308 ×27	309 ×27
# 7	300 ×28	301 ×28	302 ×28	303 ×28	304 ×28
# 8	305 ×29	306 ×29	307 ×29	308 ×29	309 ×29

Grade 6 Chapter 2 Lesson 4
Multiplication Numbers 4
Example

$$\begin{array}{r} 404 \\ \underline{\times\ 35} \\ 2020 \\ +\underline{12120} \\ 14140 \end{array}$$

1) 35 = 30 + 5

2) $\begin{array}{r} 404 \\ \underline{\times 5} \\ 2020 \end{array}$

3) $\begin{array}{r} 404 \\ \underline{\times 30} \\ 12120 \end{array}$

Review 2 A
Multiply

	a	b	c	d	e
# 1	400 × 32	401 × 32	402 × 32	403 × 32	404 × 32
# 2	405 × 33	406 × 33	407 × 33	408 × 33	409 × 33
# 3	400 × 34	401 × 34	402 × 34	403 × 34	404 × 34
# 4	405 × 35	406 × 35	407 × 35	408 × 35	409 × 35
# 5	400 × 36	401 × 36	402 × 36	403 × 36	404 × 36
# 6	405 × 37	406 × 37	407 × 37	408 × 37	409 × 37
# 7	400 × 38	401 × 38	402 × 38	403 × 38	404 × 38
# 8	405 × 39	406 × 39	407 × 39	408 × 39	409 × 39

Grade 6 Chapter 2 Lesson 4
Multiplication Numbers 4
Example

$$\begin{array}{r} 506 \\ \underline{\times 45} \\ 2530 \\ \underline{20240} \\ 22770 \end{array}$$

1) $45 = 40 + 5$

2) $\begin{array}{r} 506 \\ \underline{\times 5} \\ 2530 \end{array}$

3) $\begin{array}{r} 506 \\ \underline{\times 40} \\ 20240 \end{array}$

Review 2 B
Multiply

	a	b	c	d	e
# 1	500 × 42	501 × 42	502 × 42	503 × 42	504 × 42
# 2	505 × 43	506 × 43	507 × 43	508 × 43	509 × 43
# 3	500 × 44	501 × 44	502 × 44	503 × 44	504 × 44
# 4	505 × 45	506 × 45	507 × 45	508 × 45	509 × 45
# 5	500 × 46	501 × 46	502 × 46	503 × 46	504 × 46
# 6	505 × 47	506 × 47	507 × 47	508 × 47	509 × 47
# 7	500 × 48	501 × 48	502 × 48	503 × 48	504 × 48
# 8	505 × 49	506 × 49	507 × 49	508 × 49	509 × 49

Grade 6 Chapter 2 Lesson 5
Multiplication Numbers 5
Example

604 × 55 = 33220	1) 55 = 50 + 5	
	2) 604 × 5 = 3020	3) 604 × 50 = 30200

$$\begin{array}{r} 604 \\ \times\ 55 \\ \hline 3020 \\ +30200 \\ \hline 33220 \end{array}$$

Review 1 A
Multiply

	a	b	c	d	e
# 1	600 ×52	601 ×52	602 ×52	603 ×52	604 × 52
# 2	605 ×53	606 ×53	607 ×53	608 ×53	609 ×53
# 3	600 ×54	601 ×54	602 ×54	603 ×54	604 ×54
# 4	605 ×55	606 ×55	607 ×55	608 ×55	609 ×55
# 5	600 × 56	601 × 56	602 × 56	603 × 56	604 × 56
# 6	605 ×57	606 ×57	607 ×57	608 ×57	609 ×57
# 7	600 ×58	601 ×58	602 ×58	603 ×58	604 ×58
# 8	605 ×59	606 ×59	607 ×59	608 ×59	609 ×59

Grade 6 Chapter 2 Lesson 5
Multiplication Numbers 5
Example

```
   3
   2
  705      1) 65 = 60 + 5     2)  2        3)    3
 ×65                             705            705
 3525                          ×  5            ×60
+42300                           3525          42300
 45825
```

Review 1 B
Multiply

	a	b	c	d	e
# 1	700 ×62	701 ×62	702 ×62	703 ×62	704 × 62
# 2	705 ×63	706 ×63	707 ×63	708 ×63	709 ×63
# 3	700 ×64	701 ×64	702 ×64	703 ×64	704 ×64
# 4	705 ×65	706 ×65	707 ×65	708 ×65	709 ×65
# 5	700 × 66	701 × 66	702 × 66	703 × 66	704 × 66
# 6	705 ×67	706 ×67	707 ×67	708 ×67	709 ×67
# 7	700 ×68	701 ×68	702 ×68	703 ×68	704 ×68
# 8	705 ×69	706 ×69	707 ×69	708 ×69	709 ×69

Grade 6 Chapter 2 Lesson 5
Multiplication Numbers 5
Example

```
   2
   1
  803      1) 75 = 70 + 5     2)    1          3)    2
 × 75                              803               803
 4015                             ×  5              × 70
+56210                            4015             56210
 60225
```

Review 2 A
Multiply

	a	b	c	d	e
# 1	800 × 72	801 × 72	802 × 72	803 × 72	804 × 72
# 2	805 × 73	806 × 73	807 × 73	808 × 73	809 × 73
# 3	800 × 74	801 × 74	802 × 74	803 × 74	804 × 74
# 4	805 × 75	806 × 75	807 × 75	808 × 75	809 × 75
# 5	800 × 76	801 × 76	802 × 76	803 × 76	804 × 76
# 6	805 × 77	806 × 77	807 × 77	808 × 77	809 × 77
# 7	800 × 78	801 × 78	802 × 78	803 × 78	804 × 78
# 8	805 × 79	806 × 79	807 × 79	808 × 79	809 × 79

Grade 6 Chapter 2 Lesson 5
Multiplication Numbers 5
Example

$$\begin{array}{r} {}^{3} \\ {}^{2} \\ 904 \\ \times\ 85 \\ \hline 4520 \\ +72320 \\ \hline 76840 \end{array}$$

1) 85 = 80 + 5

2) $\begin{array}{r} {}^{2} \\ 904 \\ \times\ 5 \\ \hline 4520 \end{array}$ 3) $\begin{array}{r} {}^{3} \\ 904 \\ \times\ 80 \\ \hline 72320 \end{array}$

Review 2 B
Multiply

	a	b	c	d	e
# 1	900 ×82	901 ×82	902 ×82	903 ×82	904 ×82
# 2	905 ×83	906 ×83	907 ×83	908 ×83	909 ×83
# 3	900 ×84	901 ×84	902 ×84	903 ×84	904 ×84
# 4	905 ×85	906 ×85	907 ×85	908 ×85	909 ×85
# 5	900 ×86	901 ×86	902 ×86	903 ×86	904 ×86
# 6	905 ×87	906 ×87	907 ×87	908 ×87	909 ×87
# 7	900 ×88	901 ×88	902 ×88	903 ×88	904 ×88
# 8	905 ×89	906 ×89	907 ×89	908 ×89	909 ×89

Grade 6 Chapter 2 Lesson 6
Multiplication numbers 6
Example

Factor × Factor = Product   2 × 3 = 6

14005	1)	14005=10000 + 4000 + 5
× 6	2)	6 × 5 = 30 (Product 1 = P 1)
30 (P1)	3)	6×4000 = 24000 (Product 2 = P 2)
24000 (P2)	4)	6×10000 = 60000 (Product 3 = P 3)
60000 (P3)	5)	60000 + 24000 + 30 = 84030 (Product = P)
84030 (P)		

Review 1 A
Multiply numbers

	a	b	c	d	e
1)	12000 × 8	12001 × 8	12002 × 8	12003 × 8	12004 × 8
2)	12005 × 8	12006 × 8	12007 × 8	12008 × 8	12009 × 8
3)	13000 × 7	13001 × 7	13002 × 7	13003 × 7	13004 × 7
4)	13005 × 7	13006 × 7	13007 × 7	13008 × 7	13009 × 7
5)	14000 × 6	14001 × 6	14002 × 6	14003 × 6	14004 × 6
6)	14005 × 6	14006 × 6	14007 × 6	14008 × 6	14009 × 6
7)	15000 × 5	15001 × 5	15002 × 5	15003 × 5	15004 × 5
8)	15005 × 5	15006 × 5	15007 × 5	15008 × 5	15009 × 5

Grade 6 Chapter 2 Lesson 6
Multiplication numbers 6
Example

Top 5 3 19005 × 6 114030(P)	1) 6× 5 = 30 2) Put 0 below 6 (P) 3) Put 3 above 0(top) 4) 6 × 0 = 0 5) 0 + 3 = 3 6) Put 3 ahead 0 (P) 7) 6 × 0 = 0	8) Put 0 ahead 3 (P) 9) 6 × 9 = 54 10) Put 4 ahead 0 (P) 11) Put 5 above 1(top) 12) 6 × 1 = 6 13) 6 + 5 = 11 14) Put 11 ahead 4(P)

Review 1 B
Multiply numbers

	a	b	c	d	e
1)	16000 × 7	16001 × 7	16002 × 7	16003 × 7	16004 × 7
2)	16005 × 7	16006 × 7	16007 × 7	16008 × 7	16009 × 7
3)	17000 × 8	17001 × 8	17002 × 8	17003 × 8	17004 × 8
4)	17005 × 9	17006 × 9	17007 × 9	17008 × 9	17009 × 9
5)	18000 × 6	18001 × 6	18002 × 6	17003 × 6	17004 × 6
6)	17005 × 7	17006 × 7	17007 × 7	17008 × 7	17009 × 7
7)	18000 × 8	18001 × 8	18002 × 8	18003 × 8	18004 × 8
8)	18005 × 9	18006 × 9	18007 × 9	18008 × 9	18009 × 9

Grade 6 Chapter 2 Lesson 6
Multiplication numbers 6
Factor × Factor = Product   2 × 3 = 6

Example

```
   1004        1)  15 = 10 + 5
 ×  15         2)  1004 ×10 =10040 (Product 2 = P2)
  5020 (P1)    3)  1004
+10040 (P2)          ×5
 15060 (P)          5020 (Product 1 = P1)
               4)  5020 + 10040 = 15060 (Product = P)
```

Review 2 A

Multiply numbers

	a	b	c	d	e
1)	1000 ×12	1001 ×12	1002 ×12	1003 ×12	1004 × 12
2)	1005 ×13	1006 ×13	1007 ×13	1008 ×13	1009 ×13
3)	1000 ×14	1001 ×14	1002 ×14	1003 ×14	1004 ×14
4)	1005 ×15	1006 ×15	1007 ×15	1008 ×15	1009 ×15
5)	1000 × 16	1001 × 16	1002 × 16	1003 × 16	1004 × 16
6)	1005 ×17	1006 ×17	1007 ×17	1008 ×17	1009 ×17
7)	1000 ×18	1001 ×18	1002 ×18	1003 ×18	1004 ×18
8)	1005 ×19	1006 ×19	1007 ×19	1008 ×19	1009 ×19

Grade 6 Chapter 2 Lesson 6
Multiplication numbers 6
Example

6304	1) 15 = 10 + 5	
× 15	2) 6304	3) 6304
31520 (P1)	× 5	× 10
+63040 (P2)	31520 (P1)	63040 (P2)
94560 (P)	4) 31520 +63040 = 94560 (Product=P)	

Review 2 B
Multiply Numbers

	a	b	c	d	e
1)	3000 × 17	3001 × 17	3002 × 17	3003 × 17	3004 × 17
2)	3005 × 18	3006 × 18	3007 × 18	3008 × 18	3009 × 18
3)	4000 × 19	4001 × 19	4002 × 19	4003 × 19	4004 × 19
4)	4005 × 15	4006 × 15	4007 × 15	4008 × 15	4009 × 15
5)	5000 × 16	5001 × 16	5002 × 16	5003 × 16	5004 × 16
6)	5005 × 17	5006 × 17	5007 × 17	5008 × 17	5009 × 17
7)	6000 × 18	6001 × 18	6002 × 18	6003 × 18	6004 × 18
8)	6005 × 19	6006 × 19	6007 × 19	6008 × 19	6009 × 19

Grade 6 Chapter 2 Lesson 7
Multiplication numbers 7
Example

405	1) 106 = 100 + 6	
×106	2) 405	3) 405
2430 (P1)	× 6	×100
+40500 (P2)	2430 (P1)	40500 (P2)
42930 (P)		

Review 1 A
Multiply numbers

	a	b	c	d	e
1)	300 ×107	301 ×107	302 ×107	303 ×107	304 ×107
2)	305 ×108	306 ×108	307 ×108	308 ×108	309 ×108
3)	400 ×109	401 ×109	402 ×109	403 ×109	404 ×109
4)	405 ×105	406 ×105	407 ×105	408 ×105	409 ×105
5)	500 ×106	501 ×106	502 ×106	503 ×106	504 ×106
6)	505 ×107	506 ×107	507 ×107	508 ×107	509 ×107
7)	600 ×108	601 ×108	602 ×108	603 ×108	604 ×108
8)	605 ×109	606 ×109	607 ×109	608 ×109	609 ×109

Grade 6 Chapter 2 Lesson 7
Multiplication numbers 7
Example 1

15	2004
×2004	× 15
60	10020
0000	+20040
30000	30060
30060	(easy)

Example 2

16	2005
×2005	× 16
80	12030
0000	+20050
32000	32080
32080	(easy)

Review 1 B
Multiply numbers

	a	b	c	d	e
1)	12× 2000	12×2001	12×2002	12×2003	12×2004
2)	13×2005	13×2006	13×2007	13×2008	13×2009
3)	14×2000	14×2001	14×2002	14×2003	14×2004
4)	15×2005	15×2006	15×2007	15×2008	15×2009
5)	16× 2000	16×2001	16×2002	16×2003	16×2004
6)	17×2005	17×2006	17×2007	17×2008	17×2009
7)	18×2000	18×2001	18×2002	18×2003	18×2004
8)	19×2005	19×2006	19×2007	19×2008	19×2009

Grade 6 Chapter 2 Lesson 7
Multiplication numbers 7
Example 1

6304	15
× 15	×6304
31520	60
+63040	4500
94560	90000
(easy)	94560

Example 2

6408	25
× 25	×6408
32040	200
+128160	10000
160200	+150000
(easy)	160200

Review 2 A
Multiply Numbers

	a	b	c	d	e
1)	15× 3000	15×3001	15×3002	15×3003	15×3004
2)	16× 3005	16×3006	16×3007	16×3008	16×3009
3)	17× 4000	17×4001	17×4002	17×4003	17×4004
4)	18× 4005	18×4006	18×4007	18×4008	18×4009
5)	19× 5000	19×5001	19×5002	19×5003	19×5004
6)	15× 5005	15×5006	15×5007	15×5008	15×5009
7)	16× 6000	16×6001	16×6002	16×6003	16×6004
8)	17× 6005	17×6006	17×6007	17×6008	17×6009

Grade 6 Chapter 2 Lesson 7
Multiplication numbers 7

Example 1

16000	16000
× 5	× 5
80000	80000

Example 2

18000	18000
× 5	× 5
90000	90000

Review 2 B

Multiply Numbers

	a	b	c	d	e
1)	5× 13000	5×13001	5×13002	5×13003	5×13004
2)	6× 13005	6×13006	6×13007	6×13008	6×13009
3)	7× 14000	7×14001	7×14002	7×14003	7×14004
4)	8× 14005	8×14006	8×14007	8×14008	8×14009
5)	9× 15000	9×15001	9×15002	9×15003	9×15004
6)	5× 15005	5×15006	5×15007	5×15008	5×15009
7)	6× 16000	6×16001	6×16002	6×16003	6×16004
8)	7× 16005	7×16006	7×16007	7×16008	7×16009

Grade 6 Chapter 2 Lesson 8

Solve examples yourself then check your solutions

Examples 1

Multiply Numbers

1) 1056	2) 2056	3) 3506	4) 4506	5) 5506
× 5	× 5	× 5	× 5	× 5
30	30	30	30	30
250	250	2500	2500	2500
+5000	+10000	+15000	+20000	+ 25000
5280	10280	17530	22530	27530

6) 6704	7) 7406	8) 8406	9) 9406	10) 9046
× 5	× 5	× 5	× 5	× 5
20	30	30	30	30
3500	2000	2000	2000	200
+30000	+35000	+40000	+45000	+45000
33520	37030	42030	47030	47230

11) 104	12) 204	13) 304	14) 404	15) 504
×15	×15	×15	×15	×15
520	1020	1520	2020	2520
1040	2040	3040	4040	5040
1560	3060	4560	6060	7560

16) 604	17) 704	18) 804	19) 904	20) 25
×25	×25	×25	×25	× 904
3020	3520	4020	4520	100
+12080	+14080	+16080	+ 18080	+22500
15100	17600	20100	22600	22600

Grade 6 Chapter 2 Lesson 8

Solve examples yourself then check your solutions

Examples 2

Multiply Numbers

1)	2)	3)	4)	5)
12001	13002	23004	34005	56007
× 9	× 8	× 7	× 6	× 5
108009	104016	161028	204030	280035

6)	7)	8)	9)	10)
67004	34005	45006	56007	67008
× 9	× 8	× 7	× 6	× 5
603036	272040	315042	336042	335040

11)	12)	13)	14)	15)
604	704	804	904	205
×205	×205	× 205	×205	× 904
3020	3520	4020	4520	820
+120800	+140800	+160800	+180800	+184500
123820	144320	164820	185320	185320

16)	17)	18)	19)	20)
1200	2304	3405	4506	5607
× 15	× 25	× 35	×45	× 65
6000	11520	17025	22530	28035
+12000	+46080	+102150	+ 180240	+336420
18000	57600	119175	202770	364455

Grade 6 Chapter 2 Lesson 9

Examples

1)	2)	3)
607	708	809
×708	×809	×908
4856	6372	6472
+424900	+566400	+728100
429756	572772	734572

Review 2 A for Test 2 A

Multiply Number

1 1023×9

2 5067×5

3 908×15

4 504×65

5 10203×9

6 50607×5

7 1203×95

8 5607×45

9 102×203

10 607×708

Grade 6 Chapter 2 Lesson 9

1) $\begin{array}{r} 908 \\ \underline{\times 807} \\ 6356 \\ \underline{+726400} \\ 732756 \end{array}$

2) $\begin{array}{r} 807 \\ \underline{\times 706} \\ 4842 \\ \underline{+564900} \\ 569742 \end{array}$

Review 2 B for Test 2 B

1 2034×8

2 6078×4

3 807×25

4 403×75

5 20304×8

6 60708×4

7 2304×85

8 6708×35

9 203×304

10 708×809

Grade 6 Chapter 2 Lesson 10

Test 2 A

Multiply Number

1 3045×7

2 7089×3

3 706×35

4 302×85

5 30405×7

6 70809×3

7 3405×75

8 7809×25

9 304×506

10 908×807

Grade 6 Chapter 2 Lesson 10

Test 2 B

Multiply Numbers

1 4056×6

2 8097×2

3 605×45

4 201×95

5 40506×6

6 80907×2

7 4506×65

8 8907×15

9 506×607

10 807×706

Grade 6 Answers Chapter 2

Lesson 1

Review 1A

	a	b	c	d	e
1)	20	22	24	26	28
2)	30	33	36	39	42
3)	40	44	48	52	56
4)	50	55	60	65	70
5)	60	66	72	78	84
6)	70	77	84	91	98
7)	80	88	96	104	112
8)	90	99	108	117	126

Review 1 B

	a	b	c	d	e
1)	30	32	34	36	38
2)	45	48	51	54	57
3)	60	64	68	72	76
4)	75	80	85	90	95
5)	90	96	102	108	114
6)	105	112	119	126	133
7)	120	128	136	144	152
8)	135	144	153	162	171

Lesson 1 Review 2 A

	a	b	c	d	e
# 1	2040	2044	2048	2052	2056
# 2	3093	3099	3105	3111	3117
# 3	4160	4168	4176	4184	4192
# 4	5255	5265	5275	5285	5295
# 5	9600	9612	9624	9636	9648
# 6	11907	11921	11935	11949	11963
# 7	14400	14416	14432	14448	14464
# 8	17109	17127	17145	17163	17181

Lesson 1 Review 2 B

	a	b	c	d	e
# 1	4040	4044	4048	4052	4056
# 2	6093	6099	6105	6111	6117
# 3	8160	8168	8176	8184	8192
# 4	10255	10265	10275	10285	10295
# 5	15600	15612	15624	15636	15648
# 6	18907	18921	18935	18949	18963
# 7	22400	22416	22432	22448	22464
# 8	26109	26127	26145	26163	26181

Lesson 2 Review 1 A Grade 6 Answers Chapter 2

	a	b	c	d	e
# 1	6040	6044	6048	6052	6056
# 2	9093	9099	9105	9111	9117
# 3	12160	12168	12176	12184	12192
# 4	15255	15265	15275	15285	15295
# 5	21600	21612	21624	21636	21648
# 6	25907	25921	25935	25949	25963
# 7	30400	30416	30432	30448	30464
# 8	35109	35127	35145	35163	35181

Lesson 2 Review 1 B

	a	b	c	d	e
# 1	8040	8044	8048	8052	8056
# 2	12093	12099	12105	12111	12117
# 3	16160	16168	16176	16184	16192
# 4	20255	20265	20275	20285	20295
# 5	27600	27612	27624	27636	27648
# 6	32907	32921	32935	32949	32963
# 7	38400	38416	38432	38448	38464
# 8	44109	44127	44145	44163	44181

Lesson 2 Review 2 A

	a	b	c	d	e
# 1	10040	10044	10048	10052	10056
# 2	15093	15099	15105	15111	15117
# 3	20160	20168	20176	20184	20192
# 4	25255	25265	25275	25285	25295
# 5	33600	33612	33624	33636	33648
# 6	39907	33921	39935	39949	39963
# 7	46400	46416	46432	46448	46464
# 8	53109	53127	53145	53163	53181

Lesson 2 Review 2 B

	a	b	c	d	e
# 1	12040	12044	12048	12052	12056
# 2	18093	18099	18105	18111	18117
# 3	24160	24168	24176	24184	24192
# 4	30255	30265	30275	30285	30295
# 5	39600	39612	39624	39636	39648
# 6	46907	46921	46935	46949	46963
# 7	54400	54416	54432	54448	54464
# 8	62109	62127	62145	62163	62181

Lesson 3 Review 1 A

	a	b	c	d	e
# 1	14040	14044	14048	14052	14056
# 2	21093	21099	21105	21111	21117
# 3	28160	28168	28176	28184	28192
# 4	35255	35265	35275	35285	35295
# 5	45600	45612	45624	45636	45648
# 6	53907	53921	53935	53949	53963
# 7	62400	62416	62432	62448	62464
# 8	71109	71127	71145	71163	71181

Lesson 3 Review 1 B

	a	b	c	d	e
# 1	16040	16044	16048	16052	16056
# 2	24093	24099	24105	24111	24117
# 3	32160	32168	32176	32184	32192
# 4	40255	40265	40275	40285	40295
# 5	51600	51612	51624	51636	51648
# 6	60907	60921	60935	60949	60963
# 7	70400	70416	70432	70448	70464
# 8	80109	80127	80145	80163	80181

Lesson 3 Review 2 A

	a	b	c	d	e
# 1	18040	18044	18048	18052	18056
# 2	27093	27099	27105	27111	27117
# 3	36160	36168	36176	36184	36192
# 4	45255	45265	45275	45285	45295
# 5	57600	57612	57624	57636	57648
# 6	67907	67921	67935	67949	67963
# 7	78400	78416	78432	78448	78464
# 8	89109	89127	89145	89163	89181

Lesson 3 Review 2 B

	a	b	c	d	e
# 1	1200	1212	1224	1236	1248
# 2	1365	1378	1391	1404	1417
# 3	1400	1414	1428	1442	1456
# 4	1575	1590	1605	1620	1635
# 5	1600	1616	1632	1648	1664
# 6	1785	1802	1819	1836	1853
# 7	1800	1818	1836	1854	1872
# 8	1995	2014	2033	2952	2071

Lesson 4 Review 1 A

	a	b	c	d	e
# 1	2400	2412	2424	2436	2448
# 2	2665	2678	2691	2704	2717
# 3	2800	2814	2828	2842	2856
# 4	3075	3090	3105	3120	3135
# 5	3200	3216	3232	3248	3264
# 6	3485	3502	3519	3536	3553
# 7	3600	3618	3636	3654	3672
# 8	3895	3914	3933	3952	3971

Lesson 4 Review 1 B

	a	b	c	d	e
# 1	6600	6622	6644	6666	6688
# 2	7015	7038	7061	7084	7107
# 3	7200	7224	7248	7272	7296
# 4	7625	7650	7675	7700	7725
# 5	7800	7826	7852	7878	7904
# 6	8235	8262	8289	8316	8343
# 7	8400	8428	8456	8484	8512
# 8	8845	8874	8903	8932	8961

Lesson 4 Review 2 A

	a	b	c	d	e
# 1	12800	12832	12864	12896	12928
# 2	13365	13398	13431	13464	13497
# 3	13600	13634	13668	13702	13736
# 4	14175	14210	14245	14280	14315
# 5	14400	14436	14472	14508	14544
# 6	14985	15022	15059	15096	15133
# 7	15200	15238	15276	15314	15352
# 8	15795	15834	15873	15912	15951

Lesson 4 Review 2 B

	a	b	c	d	e
# 1	21000	21042	21084	21126	21168
# 2	21715	21758	21801	21844	21887
# 3	22000	22044	22088	22132	22176
# 4	22725	22770	22815	22860	22905
# 5	23000	23046	23092	23138	23184
# 6	23735	23782	23829	23876	23923
# 7	24000	24048	24096	24144	24192
# 8	24745	24794	24843	24892	24941

Lesson 5 Review 1 A

	a	b	c	d	e
# 1	31200	31252	31304	31356	31408
# 2	32065	32118	32171	32224	32277
# 3	32400	32454	32508	32562	32616
# 4	33275	33330	33385	33440	33495
# 5	33600	33656	33712	33768	33824
# 6	34485	34542	34599	34656	34713
# 7	34800	34858	34916	34974	35032
# 8	35695	35754	35813	35872	35931

Lesson 5 Review 1 B

	a	b	c	d	e
# 1	43400	43462	43524	43586	43648
# 2	44415	44478	44541	44604	44667
# 3	44800	44864	44928	44992	45056
# 4	45825	45890	45955	46020	46085
# 5	46200	46266	46332	46398	46464
# 6	47235	47302	47369	47436	47503
# 7	47600	47668	47736	47804	47872
# 8	48645	48714	48783	48852	48921

Lesson 5 Review 2 A

	a	b	c	d	e
# 1	57600	57672	57744	57816	57888
# 2	58765	58838	58911	58984	59057
# 3	59200	59274	59348	59422	59496
# 4	60375	60450	60525	60600	60675
# 5	60800	60876	60952	61028	61104
# 6	61985	62062	62139	62216	62293
# 7	62400	62478	62556	62634	62712
# 8	63595	63674	63753	63832	63911

Lesson 5 Review 2 B

	a	b	c	d	e
# 1	73800	73882	73964	74046	74128
# 2	75115	75198	75281	75364	75447
# 3	75600	75684	75768	75852	75936
# 4	76925	77010	77095	77180	77265
# 5	77400	77486	77572	77658	77744
# 6	78735	78822	78909	78996	79083
# 7	79200	79288	79376	79464	79552
# 8	80545	80634	80723	80812	80901

Lesson 6 Review 1 A

	a	b	c	d	e
# 1	96000	96008	96016	96024	96032
# 2	96040	96048	96056	96064	96072
# 3	91000	91007	91014	91021	91028
# 4	91035	91042	91049	91056	91063
# 5	84000	84006	84012	84018	84024
# 6	84030	84036	84042	84048	84054
# 7	75000	75005	75010	75015	75020
# 8	75025	75030	75035	75040	75045

Lesson 6 Review 1 B

	a	b	c	d	e
# 1	112000	112007	112014	112021	112028
# 2	112035	112042	112049	112056	112063
# 3	136000	136008	136016	136024	136032
# 4	153045	153054	153063	153072	153081
# 5	108000	108006	108012	108018	108024
# 6	119035	119042	119049	119056	119063
# 7	144000	144008	144016	144024	144032
# 8	162045	162054	162063	162072	162081

Lesson 6 Review 2 A

	a	b	c	d	e
1)	12000	12012	12024	12036	12048
2)	13065	13078	13091	13104	13117
3)	14000	14014	14028	14042	14056
4)	15075	15090	15105	15120	15135
5)	16000	16016	16032	16048	16064
6)	17085	17102	17119	17136	17153
7)	18000	18018	18036	18054	18072
8)	19095	19114	19133	19152	19171

Lesson 6 Review 2 B

	a	b	c	d	e
1)	51000	51017	51034	51051	51068
2)	54090	54108	54126	54144	54162
3)	76000	76019	76038	76057	76076
4)	60075	60090	60105	60120	60135
5)	80000	80016	80032	80048	80064
6)	85085	85102	85119	85136	85153
7)	108000	108018	108036	108054	108072
8)	114095	114114	114133	114152	114171

Lesson 7 Review 1 A

	a	b	c	d	e
1)	32100	32207	32314	32421	32528
2)	32940	33048	33156	33264	33372
3)	43600	43709	43818	43927	44036
4)	42525	42630	42735	42840	42945
5)	53000	53106	53212	53318	53424
6)	54035	54142	54249	54356	54463
7)	64800	64908	65016	65124	65232
8)	65945	66054	66163	66272	66381

Lesson 7 Review 1 B

1)	24000	24012	24024	24036	24048
2)	26065	26078	26091	26104	26117
3)	28000	28014	28028	28042	28056
4)	30075	30090	30105	30120	30135
5)	32000	32016	32032	32048	32064
6)	34085	34102	34119	34136	34153
7)	36000	36018	36036	36054	36072
8)	38095	38114	38133	38152	38171

Lesson 7 Review 2 A

1)	45000	45015	45030	45045	45060
2)	48080	48096	48112	48128	48144
3)	68000	68017	68034	68051	68068
4)	72090	72108	72126	72144	72162
5)	95000	95019	95038	95057	95076
6)	75075	75090	75105	75120	75135
7)	96000	96016	96032	96048	96064
8)	102085	102102	102119	102136	102153

Lesson 7 Review 2 B

1)	65000	65005	65010	65015	65020
2)	78030	78036	78042	78048	78054
3)	98000	98007	98014	98021	98028
4)	112040	112048	112056	112064	112072
5)	135000	135009	135018	135027	135036
6)	75025	75030	75035	75040	75045
7)	96000	96006	96012	96018	96024
8)	112035	112042	112049	112056	112063

Grade 6 Answers Chapter 2

Lesson 9 Review 2 A

1 9207 # 2 25335 # 3 13620 # 4 32760 # 5 91827

6 253035 # 7 114285 # 8 252315 # 9 20706 # 10 429756

Lesson 9 Review 2 B

1 16272 # 2 24312 # 3 20175 # 4 30225 # 5 162432

6 242832 # 7 195840 # 8 234780 # 9 61712 # 10 572772

Lesson 10

Test 2 A

1 21315 # 2 21267 # 3 24710 # 4 25670 # 5 212835

6 212427 # 7 255375 # 8 195225 # 9 153824 # 10 732756

Test 2 B

1 24336 # 2 16194 # 3 27225 # 4 19095 # 5 243036

6 161814 # 7 292890 # 8 133605 # 9 307142 # 10 569742

Grade 6

Chapter 3

Division

Numbers 3

Grade 6 Chapter 3 Contents
Division Numbers
Lesson 1 – 10
Basic Problems

Work with lesson 1, read rules and examples, 2) Solve the problems from contents, 3) Check answers in the end of chapter 1, 4) Work with the next lessons 2, 3, 4, 5, …, 5) Solve Test 1 A, 6) Grading Scale: A =10 points, B=9points, C=8 points, D=7 points, F=0-6 points, 7) If you solve 10 problems correctly, you get an A, 8) If have an A, you can move on to next chapter, 9) If your grade is lower than an A, work with the mistakes, then solve Test 1B, 10) When you finish all of the chapters, solve Final Exam in last chapter.

Grade 6 Chapter 3 Lesson 1
Division Numbers 1

Fourth operation with numbers is division ÷ or :
Division means separated number by some equal groups

First number in division called the dividend	6
Second number in division called the divisor	2
Result of division called the quotient	3
Dividend ÷ divisor = quotient	$6 \div 2 = 3$

Properties of division

Any number divided by one is the same number	$5 \div 1 = 5$
Any number divided by the itself is one	$5 \div 5 = 1$
Zero divided by any number is zero	$0 \div 5 = 0$
Division by zero is not allowed	$5 \div 0$ $A \div 0$ None

Formulas: 1) $A \div 1 = A$ 2) $A \div A = 1$
3) $0 \div A = 0$ 4) $A \div 0$ None

Review 1 A
Divide numbers by using Table of multiplication

	a	b	c	d
1)	0 ÷ 6	7 ÷ 7	72 ÷ 8	0 ÷ 9
2)	12 ÷ 6	21 ÷ 7	64 ÷ 8	63 ÷ 9
3)	30 ÷ 6	28 ÷ 7	48 ÷ 8	45 ÷ 9
4)	42 ÷ 6	42 ÷ 7	32 ÷ 8	27 ÷ 9
5)	48 ÷ 6	63 ÷ 7	16 ÷ 8	9 ÷ 9
6)	54 ÷ 6	56 ÷ 7	8 ÷ 8	18 ÷ 9
7)	36 ÷ 6	49 ÷ 7	24 ÷ 8	36 ÷ 9
8)	24 ÷ 6	35 ÷ 7	40 ÷ 8	54 ÷ 9
9)	18 ÷ 6	14 ÷ 7	56 ÷ 8	72 ÷ 9
10)	6 ÷ 6	0 ÷ 7	0 ÷ 8	81 ÷ 9

Grade 6 Chapter 3 Lesson 1
Division Numbers 1

To check multiplication use division

2×3=6 Check 6÷2=3 or 6÷3=2

To check division use multiplication and division

6÷2=3 Check 2×3=6 or 6÷3=2

Division without remainder R = 0

```
           5 - Quotient
Divisor 7⌋ 35 – Dividend
          -35
           0 – Remainder
```

Division with remainder R > 0

```
           5 - Quotient
Divisor 7 |36 – Dividend
          -35
           1 – Remainder
```

Review 1 B

Complete the Table of Division by using Table of Multiplication (page 51)

		a	b	c	d	e	f	g	h
	×	2	?	4	?	6	?	8	?
11)	1								9
12)	?								18
13)	3							24	
14)	?							32	
15)	5						35		
16)	?					36			
17)	7				35				
18)	?			32					
19)	9		27						
20)	?	20							

Grade 6 Chapter 3 Lesson 1

Division Numbers 1

6÷2=3 dividend 6 divisor 2 quotient 3(q)

Every time bring down just one digit

Example 1

105 (q)	1)	$14 \times \underline{1} = 14$	Put 1 in quotient (q)
14 ⟌ 1470	2)	$14 - 14 = 0$	
-10	3)	Bring down the 7	
07	4)	$14 \times \underline{0} = 0$	Put 0 in quotient (q)
-0	5)	$7 - 0 = 7$	
70	6)	Bring down the 0	
- 70	7)	$14 \times \underline{5} = 70$	Put 5 in quotient(q)
0	8)	$70 - 70 = 0$	

Example 2

125 (q)	1)	$18 \times \underline{1} = 18$	Put 1 in quotient (q)
18 ⟌ 2250	2)	$22 - 18 = 4$	
-18	3)	Bring down the 5	
45	4)	$18 \times \underline{2} = 36$	Put 2 in quotient (q)
-36	5)	$45 - 36 = 9$	
90	6)	Bring down the 0	
-90	7)	$18 \times \underline{5} = 90$	Put 5 in quotient (q)
0	8)	$90 - 90 = 0$	

Review 2 A

Divide numbers

	a	b	c	d
1)	1200÷ 12	1365 ÷ 13	1526÷ 14	1560 ÷ 15
2)	1212 ÷12	1378 ÷13	1512 ÷14	1545 ÷15
3)	1224 ÷ 12	1391 ÷ 13	1498 ÷ 14	1530 ÷15
4)	1236 ÷ 12	1404 ÷ 13	1484 ÷ 14	1515 ÷ 15
5)	1248 ÷ 12	1417 ÷ 13	1470 ÷ 14	1500 ÷15

Review 2 B

Divide numbers

	a	b	c	d
6)	1680 ÷16	1768 ÷17	1800 ÷18	1995 ÷19
7)	1696 ÷ 16	1751 ÷17	1818 ÷18	2014 ÷19
8)	1712 ÷ 16	1734 ÷ 17	1836 ÷18	2033 ÷19
9)	1728 ÷16	1717 ÷17	1854 ÷18	2052 ÷19
10)	1744 ÷ 16	1700 ÷ 17	1872 ÷18	2071 ÷19

Grade 6 Chapter 3 Lesson 2
Division Numbers 2

6÷2=3 dividend 6 divisor 2 quotient 3 (q)

Every time bring down just one digit

Example 1

Long division	Steps		or
206 (q) over 5 ⟌ 1030	1)	$5 \times \underline{2} = 10$	1030 ⌊ 8
- 10	2)	$10 - 10 = 0$	-10 206 (q)
03	3)	Bring down the 3	03
-0	4)	$5 \times \underline{0} = 0$	-0
30	5)	$3 - 0 = 3$	30
-30	6)	Bring down the 0	-30
0	7)	$5 \times \underline{6} = 30$	0
	8)	$30 - 30 = 0$	

Review 1 A
Divide numbers

	a	b	c	d
1)	1000÷5	1005÷5	1010÷5	1015÷5
2)	1020÷5	1025÷5	1030÷5	1035÷5
3)	1260÷6	1266÷6	1272÷6	1278÷6
4)	1284÷6	1290÷6	1296÷6	1302÷6
5)	1540÷7	1547÷7	1554÷7	1561÷7
6)	1568÷7	1575÷7	1582÷7	1589÷7
7)	1840÷8	1848÷8	1856÷8	1864÷8
8)	1872÷8	1880÷8	1888÷8	1896÷8
9)	2160÷9	2169÷9	2178÷9	2187÷9
10)	2196÷9	2205÷9	2214÷9	2223÷9

Grade 6 Chapter 3 Lesson 2
Division Numbers 2

6÷2=3 dividend 6 divisor 2 quotient 3 (q)

Every time bring down just one digit

Example 2

295 (q)	1)	$8 \times \underline{2} = 16$	or 2360 ⌊ 8
8 ⌐2360	2)	$23 - 16 = 7$	-16 295 (q)
-16	3)	Bring down the 6	76
76	4)	$8 \times \underline{9} = 72$	-72
- 72	5)	$76 - 72 = 4$	40
40	6)	Bring down the 0	-40
-40	7)	$8 \times \underline{5} = 40$	0
0	8)	$40 - 40 = 0$	

Review 1 B
Divide numbers

	a	b	c	d
1)	1250÷5	1255÷5	1260÷5	1265÷5
2)	1270÷5	1275÷5	1280÷5	1285÷5
3)	1560÷6	1566÷6	1572÷6	1578 ÷6
4)	1584÷6	1590÷6	1596÷6	1602 ÷6
5)	1890÷7	1897 ÷7	1904÷7	1911÷7
6)	1918÷7	1925÷7	1932 ÷7	1939 ÷7
7)	2240÷8	2248÷8	2256÷8	2264 ÷8
8)	2272÷8	2280÷8	2288÷8	2296÷8
9)	2610÷9	2619÷9	2628÷9	2637÷9
10)	2646÷9	2655÷9	2664÷9	2673÷9

Grade 6 Chapter 3 Lesson 2
Division Numbers 2

6÷2=3 dividend 6 divisor 2 quotient 3 (q)

Every time bring down just one digit

Example

```
  305 (q)
5|1525
 -15
  02
  -0
   25
  -25
    0
```

1) $5 \times \underline{3} = 15$ Put 3 in quotient (q)
2) $15 - 15 = 0$
3) Bring down the 2
4) $5 \times \underline{0} = 0$ Put 0 in quotient (q)
5) $2 - 0 = 2$
6) Bring down the 5
7) $5 \times \underline{5} = 25$ Put 5 in quotient (q)
8) $25 - 25 = 0$

Review 2 A
Divide numbers

	a	b	c	d
1)	1500÷5	1505÷5	1510÷5	1515÷5
2)	1520÷5	1525÷5	1530÷5	1535÷5
3)	1860÷6	1866÷6	1872÷6	1878÷6
4)	1884÷6	1890÷6	1896÷6	1902÷6
5)	2240÷7	2247÷7	2254÷7	2261÷7
6)	2268÷7	2275÷7	2282÷7	2289÷7
7)	2640÷8	2648÷8	2656÷8	2664÷8
8)	2672÷8	2680÷8	2688÷8	2696÷8
9)	3060÷9	3069÷9	3078÷9	3087÷9
10)	3096÷9	3105÷9	3114÷9	3123÷9

Grade 6 Chapter 3 Lesson 2
Division Numbers 2

6÷2=3 dividend 6 divisor 2 quotient 3 (q)

Every time bring down just one digit

Example

```
   395 (q)     1)  8 × 3 = 24      Put 3 in quotient (q)
8 | 3160       2)  31 – 24 = 7
  -24          3)  Bring down the 6
   76          4)  8 × 9 = 72      Put 9 in quotient (q)
  - 72         5)  76 – 72 = 4
     40        6)  Bring down the 0
    -40        7)  8×5 = 40        Put 5 in quotient (q)
      0        8)  40 – 40 = 0
```

Review 2 B
Divide Numbers

	a	b	c	d
1)	1750÷5	1755÷5	1760 ÷5	1765 ÷5
2)	1770÷5	1775÷5	1780÷5	1785÷5
3)	2160÷6	2166÷6	2172÷6	2178÷6
4)	2184 ÷6	2190÷6	2196÷6	2202÷6
5)	2590÷7	2597÷7	2604÷7	2611÷7
6)	2618÷7	2625÷7	2632÷7	2639÷7
7)	3040÷8	3048÷8	3056÷8	3064÷8
8)	3072÷8	3080÷8	3088÷8	3096÷8
9)	3510 ÷9	3519÷9	3528÷9	3537÷9
10)	3546÷9	3555÷9	3564÷9	3573÷9

Grade 6 Chapter 3 Lesson 3
Division Numbers 3

6÷2=3 dividend 6 divisor 2 quotient 3 (q)

Every time bring down just one digit

Example

407 (q)	1)	5 × <u>4</u> = 20	Put 4 in quotient (q)
5 ⟌ 2035	2)	20 – 20 = 0	
-20	3)	Bring down the 3	
03	4)	5 × <u>0</u> = 0	Put 0 in quotient (q)
- 0	5)	3 – 0 = 3	
35	6)	Bring down the 5	
-35	7)	5×<u>7</u> = 35	Put 7 in quotient (q)
0	8)	35 – 35 = 0	

Review 1 A
Divide Numbers

	a	b	c	d
1)	2000÷5	2005÷5	2010÷5	2015÷5
2)	2020÷5	2025÷5	2030÷5	2035÷5
3)	2460÷6	2466÷6	2472÷6	2478÷6
4)	2484÷6	2490÷6	2496 ÷6	2502÷6
5)	2940÷7	2947÷7	2954÷7	2961÷7
6)	2968÷7	2975÷7	2982÷7	2989 ÷7
7)	3440÷8	3448÷8	3456÷8	3464÷8
8)	3472÷8	3480÷8	3488÷8	3496÷ 8
9)	3960÷9	3969÷9	3978÷9	3987÷9
10)	3996÷9	4005÷9	4014÷9	4023÷9

Grade 6 Chapter 3 Lesson 3
Division Numbers 3

6÷2=3 dividend 6 divisor 2 quotient 3 (q)

Every time bring down just one digit

Example

```
   495
8 |3960
 -32
   76
 - 72
    40
   -40
     0
```

1) $8 \times \underline{4} = 32$ Put 4 in quotient (q)
2) $39 - 32 = 7$
3) Bring down the 6
4) $8 \times \underline{9} = 72$ Put 9 in quotient (q)
5) $76 - 72 = 4$
6) Bring down the 0
7) $8 \times \underline{5} = 40$ Put 5 in quotient (q)
8) $40 - 40 = 0$

Review 1 B
Divide Numbers

	a	b	c	d
1)	2250÷5	2255÷5	2260÷5	2265÷5
2)	2270÷5	2275÷5	2280÷5	2285÷5
3)	2760÷6	2766÷6	2772 ÷6	2778÷6
4)	2784 ÷6	2790÷6	2796÷6	2802÷6
5)	3290÷7	3297÷7	3304÷7	3311÷7
6)	3318÷7	3325÷7	3332÷7	3339÷7
7)	3840÷8	3848÷8	3856÷8	3864÷8
8)	3872÷8	3880÷8	3888÷8	3896÷8
9)	4410÷9	4419÷9	4428÷9	4437÷9
10)	4446÷9	4455÷9	4464÷9	4473÷9

Grade 6 Chapter 3 Lesson 3
Division Numbers 3

6÷2=3 dividend 6 divisor 2 quotient 3 (q)

Every time bring down just one digit

Example

$\underline{\ \ 508}$ (q)	1)	$5 \times \underline{5} = 25$	Put 5 in quotient (q)
5 ⟌ 2540	2)	$25 - 25 = 0$	
$\underline{-25}$	3)	Bring down the 4	
04	4)	$5 \times \underline{0} = 0$	Put 0 in quotient (q)
$\underline{-0}$	5)	$4 - 0 = 4$	
40	6)	Bring down the 0	
$\underline{-40}$	7)	$5 \times \underline{8} = 40$	Put 8 in quotient (q)
0	8)	$40 - 40 = 0$	

Review 2 A
Divide Numbers

	a	b	c	d
1)	2500÷5	2505÷5	2510÷5	2515÷5
2)	2520÷5	2525÷5	2530÷5	2535÷5
3)	3060÷6	3066÷6	3072÷6	3078÷6
4)	3084÷6	3090÷6	3096÷6	3102÷ 6
5)	3640÷7	3647÷7	3654÷7	3661÷7
6)	3668÷ 7	3675÷7	3682÷7	3689÷7
7)	4240÷8	4248÷8	4256÷8	4264÷8
8)	4272÷8	4280÷8	4288÷8	4296÷8
9)	4860÷9	4869÷9	4878÷9	4887÷9
10)	4896÷9	4905÷9	4914÷9	4923÷9

Grade 6 Chapter 3 Lesson 3
Division Numbers 3

6÷2=3 dividend 6 divisor 2 quotient 3 (q)

Every time bring down just one digit

Example

```
   595 (q)
8|4760
 -40
  76
 - 72
   40
  -40
    0
```

1) $8 \times \underline{5} = 40$ Put 5 in quotient (q)
2) $47 - 40 = 7$
3) Bring down the 6
4) $8 \times \underline{9} = 72$ Put 9 in quotient (q)
5) $76 - 72 = 4$
6) Bring down the 0
7) $8 \times \underline{5} = 40$ Put 5 in quotient (q)
8) $40 - 40 = 0$

Review 2 B
Divide Numbers

	a	b	c	d
1)	2750÷5	2755÷5	2760÷5	2765÷5
2)	2770÷5	2775÷5	2780÷5	2785÷5
3)	3360÷6	3366÷6	3372÷6	3378÷6
4)	3384÷6	3390÷6	3396÷6	3402÷ 6
5)	3990÷7	3997÷7	4004÷7	4011÷7
6)	4018÷ 7	4025÷7	4032÷7	4039÷7
7)	4640÷8	4648÷8	4656÷8	4664÷8
8)	4672÷8	4680÷8	4688÷8	4696÷8
9)	5310÷9	5319÷9	5328÷9	5337÷9
10)	5346÷9	5355÷9	5364÷9	5373÷9

Grade 6 Chapter 3 Lesson 4
Division Numbers 4

6÷2=3 dividend 6 divisor 2 quotient 3 (q)

Every time bring down just one digit. Example 1

Long division	Step		Alternative
1056 (q)	1)	5 × 1 = 5	or 5280 | 5
5 | 5280	2)	5 – 5 = 0	-5 1056 (q)
-5	3)	Bring down the 2	02
02	4)	5 × 0 = 0	-0
-0	5)	2 – 0 = 2	28
28	6)	Bring down the 8	-25
-25	7)	5×5 = 25	30
30	8)	28 – 25 = 3	- 30
-30	9)	Bring down the 0	0
0	10)	5 × 6 = 30	
	11)	30 – 30 = 0	

Review 1 A Divide numbers

	a	b	c	d
1)	2040÷ 2	3075 ÷ 3	4120 ÷ 4	5175 ÷ 5
2)	2042 ÷ 2	3078 ÷3	4124 ÷4	5180 ÷5
3)	2044 ÷ 2	3081 ÷ 3	4128 ÷ 4	5185 ÷5
4)	2046 ÷ 2	3084 ÷ 3	4132 ÷ 4	5190 ÷ 5
5)	2048 ÷ 2	3087 ÷ 3	4136 ÷ 4	5195 ÷5

Review 1 B Divide Numbers

	a	b	c	d
6)	7200 ÷ 6	8435 ÷7	10400 ÷8	11745 ÷9
7)	7206 ÷ 6	8442 ÷7	10408 ÷8	11754 ÷9
8)	7212 ÷ 6	8449 ÷7	10416 ÷8	11763÷9
9)	7218 ÷ 6	8456 ÷7	10424 ÷8	11772 ÷9
10)	7224 ÷ 6	8463 ÷7	10432 ÷8	11781 ÷9

Grade 6 Chapter 3 Lesson 4
Division Numbers 4

6÷2=3   dividend 6   divisor 2   quotient 3 (q)

Every time bring down just one digit. Example 1

2046 (quotient)	1)	$5 \times \underline{2} = 10$	Put 2 in quotient (q)
5 ⟌ 10230	2)	$10 - 10 = 0$	
-10	3)	Bring down the 2	
02	4)	$5 \times \underline{0} = 0$	Put 0 in quotient (q)
-0	5)	$2 - 0 = 2$	
23	6)	Bring down the 3	
-20	7)	$5 \times \underline{4} = 20$	Put 4 in quotient (q)
30	8)	$23 - 20 = 3$	
- 30	9)	Bring down the 0	
0	10)	$5 \times \underline{6} = 30$	Put 6 in quotient (q)
	11)	30–30=0	

Review 2 A Divide Numbers

	a	b	c	d
1)	4040÷ 2	6075 ÷ 3	8120 ÷ 4	10175 ÷ 5
2)	4042 ÷ 2	6078 ÷3	8124 ÷ 4	10180 ÷ 5
3)	4044 ÷ 2	6081 ÷ 3	8128 ÷ 4	10185 ÷ 5
4)	4046 ÷ 2	6084 ÷ 3	8132 ÷ 4	10190 ÷ 5
5)	4048 ÷ 2	6087 ÷ 3	8136 ÷ 4	10195 ÷ 5

Review 2 B Divide numbers

	a	b	c	d
6)	13200 ÷6	15435 ÷7	18400 ÷8	20745 ÷9
7)	13206 ÷ 6	15442 ÷7	18408 ÷8	20754 ÷9
8)	13212 ÷ 6	15449 ÷7	18416 ÷8	20763 ÷9
9)	13218 ÷6	15456 ÷7	18424 ÷8	20772 ÷9
10)	13224 ÷ 6	15463 ÷7	18432 ÷8	20781 ÷9

Grade 6 Chapter 3 Lesson 5
Division Numbers 5

Example 2

Division	Step	
$\underline{1506}$ (q)	1)	$5 \times \underline{1} = 5$
5 | 7530	2)	$7 - 5 = 2$
$\underline{-5}$	3)	Bring down the 5
25	4)	$5 \times \underline{5} = 25$
$\underline{-25}$	5)	$25 - 25 = 0$
03	6)	Bring down the 3
$\underline{-0}$	7)	$5 \times \underline{0} = 0$
30	8)	$3 - 0 = 3$
$\underline{-30}$	9)	Bring down the 0
	10)	$5 \times \underline{6} = 30$
	11)	$30 - 30 = 0$

or

7530	$\underline{\mid 5}$
$\underline{-5}$	1506 (q)
25	
$\underline{-25}$	
03	
$\underline{-0}$	
30	
$\underline{-30}$	
0	

Review 1 A Divide Numbers

	a	b	c	d
# 1	6040 ÷ 2	9075 ÷ 3	12120 ÷ 4	15175 ÷ 5
# 2	6042 ÷ 2	9078 ÷ 3	12124 ÷ 4	15180 ÷ 5
# 3	6044 ÷ 2	9081 ÷ 3	12128 ÷ 4	15185 ÷ 5
# 4	6046 ÷ 2	9084 ÷ 3	12132 ÷ 4	15190 ÷ 5
# 5	6048 ÷ 2	9087 ÷ 3	12136 ÷ 4	15195 ÷ 5

Review 1 B Divide Numbers

	a	b	c	d
# 6	19200 ÷ 6	22435 ÷ 7	26400 ÷ 8	29745 ÷ 9
# 7	19206 ÷ 6	22442 ÷ 7	26408 ÷ 8	29754 ÷ 9
# 8	19212 ÷ 6	22449 ÷ 7	26416 ÷ 8	29763 ÷ 9
# 9	19218 ÷ 6	22456 ÷ 7	26424 ÷ 8	29772 ÷ 9
# 10	19224 ÷ 6	22463 ÷ 7	26432 ÷ 8	29781 ÷ 9

Grade 6 Chapter 3 Lesson 5
Division Numbers 5

20280 – dividend 5 – divisor 4056 – quotient

Every time bring down just one digit.

Example 1

Long division	Step		
4056	1)	$5 \times \underline{4} = 20$	Put 4 in quotient
5 ∣ 20280	2)	$20 - 20 = 0$	
-20	3)	Bring down the 2	
02	4)	$5 \times \underline{0} = 0$	Put 0 in quotient
-0	5)	$2 - 0 = 2$	
28	6)	Bring down the 8	
-25	7)	$5 \times \underline{5} = 25$	Put 5 in quotient
30	8)	$28 - 25 = 3$	
- 30	9)	Bring down the 0	
0	10)	$5 \times \underline{6} = 30$	Put 6 in quotient
	11)	$30 - 30 = 0$	

Review 2 A
Divide Numbers

	a	b	c	d
# 1	8040÷ 2	12075 ÷ 3	16120÷ 4	20175÷5
# 2	8042 ÷ 2	12078 ÷3	16124 ÷4	20180 ÷5
# 3	8044 ÷ 2	12081 ÷ 3	16128 ÷ 4	20185 ÷5
# 4	8046 ÷ 2	12084 ÷ 3	16132 ÷ 4	20190 ÷5
# 5	8048 ÷ 2	12087 ÷ 3	16136 ÷ 4	20195 ÷5

Grade 6 Chapter 3 Lesson 5
Division Numbers 5

22530 – dividend 5 – divisor 4506 – quotient

Every time bring down just one digit

Example 2

```
   4506
5 |22530
  -20
    25
   -25
     03
     -0
      30
    - 30
       0
```

1) $5 \times \underline{4} = 20$ Put 4 in quotient
2) $22 - 20 = 2$
3) Bring down the 5
4) $5 \times \underline{5} = 25$ Put 5 in quotient
5) $25 - 25 = 0$
6) Bring down the 3
7) $5 \times \underline{0} = 0$ Put 0 in quotient
8) $3 - 0 = 3$
9) Bring down the 0
10) $5 \times \underline{6} = 30$ Put 6 in quotient
11) $30 - 30 = 0$

Review 2 B
Divide Numbers

	a	b	c	d
# 6	25200 ÷6	29435 ÷7	34400 ÷8	38745 ÷9
# 7	25206 ÷ 6	29442 ÷7	34408 ÷8	38754 ÷9
# 8	25212 ÷ 6	29449 ÷7	34416 ÷8	38763 ÷9
# 9	25218 ÷6	29456 ÷7	34424 ÷ 8	38772 ÷9
# 10	25224 ÷ 6	29463 ÷7	34432 ÷8	38781 ÷9

Grade 6 Chapter 3 Lesson 6
Division Numbers 6

25280 – dividend   5 – divisor   5056 – quotient

Every time bring down just one digit

Example

```
     5056
5 | 25280
   -25
     02
     -0
      28
     -25
       30
     - 30
        0
```

1) 5 × 5 = 25   Put 5 in quotient
2) 20 – 20 = 0
3) Bring down the 2
4) 5 × 0 = 0   Put 0 in quotient
5) 2 – 0 = 2
6) Bring down the 8
7) 5 × 5 = 25   Put 5 in quotient
8) 28 – 25 = 3
9) Bring down the 0
10) 5 × 6 = 30   Put 6 in quotient
11) 30 – 30 = 0

Review 1 A
Divide Numbers

	a	b	c	d
# 1	10040÷ 2	15075 ÷ 3	20120÷ 4	25175 ÷ 5
# 2	10042 ÷2	15078 ÷3	20124 ÷4	25180 ÷5
# 3	10044 ÷ 2	15081 ÷ 3	20128 ÷ 4	25185 ÷5
# 4	10046 ÷ 2	15084 ÷ 3	20132 ÷ 4	25190 ÷ 5
# 5	10048 ÷ 2	15087 ÷ 3	20136 ÷ 4	25195 ÷5

Grade 6 Chapter 3 Lesson 6
Division Numbers 6

27530 – dividend   5 – divisor   5506 – quotient

Every time bring down just one digit

Example

```
   5506
5|27530
 -25
   25
  -25
   03
   -0
   30
  - 30
    0
```

1)	5 × 5 = 25	Put 5 in quotient
2)	27 – 25 = 2	
3)	Bring down the 5	
4)	5 × 5 = 25	Put 5 in quotient
5)	25 – 25 = 0	
6)	Bring down the 3	
7)	5 × 0 = 0	Put 0 in quotient
8)	3 – 0 = 3	
9)	Bring down the 0	
10)	5 × 6 = 30	Put 6 in quotient
11)	30 – 30 = 0	

Review 1 B
Divide Numbers

	a	b	c	d
# 6	31200 ÷6	36435 ÷7	42400 ÷8	47745 ÷9
# 7	31206 ÷ 6	36442 ÷7	42408 ÷8	47754 ÷9
# 8	31212 ÷ 6	36449 ÷7	42416 ÷8	47763 ÷9
# 9	31218 ÷6	36456 ÷7	42424 ÷8	47772 ÷9
# 10	31224 ÷ 6	36463 ÷7	42432 ÷8	47781 ÷9

Grade 6 Chapter 3 Lesson 6
Division Numbers 6

30280 – dividend   5 – divisor   6056 – quotient

Every time bring down just one digit

Example

$5\overline{)30280}$ = 6056	1) $5 \times \underline{6} = 30$ Put 6 in quotient
-30	2) $30 - 30 = 0$
02	3) Bring down the 2
-0	4) $5 \times \underline{0} = 0$ Put 0 in quotient
28	5) $2 - 0 = 2$
-25	6) Bring down the 8
30	7) $5 \times \underline{5} = 25$ Put 5 in quotient
- 30	8) $28 - 25 = 3$
0	9) Bring down the 0
	10) $5 \times \underline{6} = 30$ Put 6 in quotient
	11) $30 - 30 = 0$

Review 2 A
Divide Numbers

	a	b	c	d
# 1	12040÷ 2	18075 ÷ 3	24120÷ 4	30175 ÷ 5
# 2	12042 ÷2	18078 ÷3	24124 ÷4	30180 ÷5
# 3	12044 ÷ 2	18081 ÷ 3	24128 ÷ 4	30185 ÷5
# 4	12046 ÷ 2	18084 ÷ 3	24132 ÷ 4	30190 ÷ 5
# 5	12048 ÷ 2	18087 ÷ 3	24136 ÷ 4	30195 ÷5

Grade 6 Chapter 3 Lesson 6
Division Numbers 6

32530 – dividend 5 – divisor 6506 – quotient

Every time bring down just one digit

Example

```
   6506
5 |32530
  -30
    25
   -25
    03
    -0
     30
   - 30
      0
```

1) $5 \times \underline{6} = 30$ Put 6 in quotient
2) $32 - 30 = 2$
3) Bring down the 5
4) $5 \times \underline{5} = 25$ Put 5 in quotient
5) $25 - 25 = 0$
6) Bring down the 3
7) $5 \times \underline{0} = 0$ Put 0 in quotient
8) $3 - 0 = 3$
9) Bring down the 0
10) $5 \times \underline{6} = 30$ Put 6 in quotient
11) $30 - 30 = 0$

Review 2 B
Divide Numbers

	a	b	c	d
# 6	37200 ÷6	43435 ÷7	50400 ÷8	56745 ÷9
# 7	37206 ÷ 6	43442 ÷7	50408 ÷8	56754 ÷9
# 8	37212 ÷ 6	43449 ÷7	50416 ÷8	56763 ÷9
# 9	37218 ÷6	43456 ÷7	50424 ÷8	56772 ÷9
# 10	37224 ÷ 6	43463 ÷7	50432 ÷8	56781 ÷9

Grade 6 Chapter 3 Lesson 7
Division Numbers 7

35280 – dividend **5 – divisor** **7056 – quotient**

Every time bring down just one digit

Example

```
   7056      1)  5 × 7 = 35       Put 7 in quotient
5 | 35280    2)  35 – 35 = 0
  -35        3)  Bring down the 2
    02       4)  5 × 0 = 0        Put 0 in quotient
    -0       5)  2 – 0 = 2
     28      6)  Bring down the 8
    -25      7)  5 × 5 = 25       Put 5 in quotient
      30     8)  28 – 25 = 3
    - 30     9)  Bring down the 0
       0     10) 5 × 6 = 30       Put 6 in quotient
             11) 30 – 30 = 0
```

Review 1 A
Divide Numbers

	a	b	c	d
# 1	14040÷ 2	21075 ÷ 3	28120÷ 4	35175 ÷ 5
# 2	14042 ÷2	21078 ÷3	28124 ÷4	35180 ÷5
# 3	14044 ÷ 2	21081 ÷ 3	28128 ÷ 4	35185 ÷5
# 4	14046 ÷ 2	21084 ÷ 3	28132 ÷ 4	35190 ÷ 5
# 5	14048 ÷ 2	21087 ÷ 3	28136 ÷ 4	35195 ÷5

Grade 6 Chapter 3 Lesson 7
Division Numbers 7

37530 – dividend **5 – divisor** **7506 – quotient**

Every time bring down just one digit

Example

```
    7506
 5|37530
  -35
    25
   -25
    03
    -0
     30
   - 30
      0
```

1)	$5 \times \underline{7} = 35$	Put 7 in quotient
2)	$37 - 35 = 2$	
3)	Bring down the 5	
4)	$5 \times \underline{5} = 25$	Put 5 in quotient
5)	$25 - 25 = 0$	
6)	Bring down the 3	
7)	$5 \times \underline{0} = 0$	Put 0 in quotient
8)	$3 - 0 = 3$	
9)	Bring down the 0	
10)	$5 \times \underline{6} = 30$	Put 6 in quotient
11)	$30 - 30 = 0$	

Review 1 B
Divide Numbers

	a	b	c	d
# 6	$43200 \div 6$	$50435 \div 7$	$58400 \div 8$	$65745 \div 9$
# 7	$43206 \div 6$	$50442 \div 7$	$58408 \div 8$	$65754 \div 9$
# 8	$43212 \div 6$	$50449 \div 7$	$58416 \div 8$	$65763 \div 9$
# 9	$43218 \div 6$	$50456 \div 7$	$58424 \div 8$	$65772 \div 9$
# 10	$43224 \div 6$	$50463 \div 7$	$58432 \div 8$	$65781 \div 9$

Grade 6 Chapter 3 Lesson 7
Division Numbers 7

40280 – dividend 5 - divisor 8056 – quotient

Every time bring down just one digit

Example

```
   8056
5 | 40280
  -40
   02
   -0
    28
   -25
     30
   - 30
      0
```

1) $5 \times \underline{8} = 40$ Put 8 in quotient
2) $40 - 40 = 0$
3) Bring down the 2
4) $5 \times \underline{0} = 0$ Put 0 in quotient
5) $2 - 0 = 2$
6) Bring down the 8
7) $5 \times \underline{5} = 25$ Put 5 in quotient
8) $28 - 25 = 3$
9) Bring down the 0
10) $5 \times \underline{6} = 30$ Put 6 in quotient
11) $30 - 30 = 0$

Review 2 A
Divide Numbers

	a	b	c	d
# 1	16040÷ 2	24075 ÷ 3	32120÷ 4	40175 ÷ 5
# 2	16042 ÷2	24078 ÷3	32124 ÷4	40180 ÷5
# 3	16044 ÷ 2	24081 ÷ 3	32128 ÷ 4	40185 ÷5
# 4	16046 ÷ 2	24084 ÷ 3	32132 ÷ 4	40190 ÷ 5
# 5	16048 ÷ 2	24087 ÷ 3	32136 ÷ 4	40195 ÷5

Grade 6 Chapter 3 Lesson 7
Division Numbers 7

42530 – dividend **5 - divisor** **8506 – quotient**

Every time bring down just one digit

Example

```
   8506
5 |42530
  -40
    25
   -25
    03
    -0
     30
   - 30
      0
```

1)	5 × <u>8</u> = 40	Put 8 in quotient
2)	42 – 40 = 2	
3)	Bring down the 5	
4)	5 × <u>5</u> = 25	Put 5 in quotient
5)	25 – 25 = 0	
6)	Bring down the 3	
7)	5 × <u>0</u> = 0	Put 0 in quotient
8)	3 – 0 = 3	
9)	Bring down the 0	
10)	5 × <u>6</u> = 30	Put 6 in quotient
11)	30 – 30 = 0	

Review 2 B
Divide Numbers

	a	b	c	d
# 6	49200 ÷6	57435 ÷7	66400 ÷8	74745 ÷9
# 7	49206 ÷ 6	57442 ÷7	66408 ÷8	74754 ÷9
# 8	49212 ÷ 6	57449 ÷7	66416 ÷8	74763 ÷9
# 9	49218 ÷6	57456 ÷7	66424 ÷8	74772 ÷9
# 10	49224 ÷ 6	57463 ÷7	66432 ÷8	74781 ÷9

Grade 6 Chapter 3 Lesson 8
Division Numbers 8

45280 – dividend   5 – divisor   9056 – quotient

Every time bring down just one digit

Example

```
    9056
5 | 45280
  -45
    02
    -0
     28
    -25
      30
    - 30
       0
```

1) $5 \times \underline{9} = 45$ Put 9 in quotient
2) $45 - 45 = 0$
3) Bring down the 2
4) $5 \times \underline{0} = 0$ Put 0 in quotient
5) $2 - 0 = 2$
6) Bring down the 8
7) $5 \times \underline{5} = 25$ Put 5 in quotient
8) $28 - 25 = 3$
9) Bring down the 0
10) $5 \times \underline{6} = 30$ Put 6 in quotient
11) $30 - 30 = 0$

Review 1 A
Divide Numbers

	a	b	c	d
# 1	18040 ÷ 2	27075 ÷ 3	36120 ÷ 4	45175 ÷ 5
# 2	18042 ÷ 2	27078 ÷ 3	36124 ÷ 4	45180 ÷ 5
# 3	18044 ÷ 2	27081 ÷ 3	36128 ÷ 4	45185 ÷ 5
# 4	18046 ÷ 2	27084 ÷ 3	36132 ÷ 4	45190 ÷ 5
# 5	18048 ÷ 2	27087 ÷ 3	36136 ÷ 4	45195 ÷ 5

Grade 6 Chapter 3 Lesson 8
Division Numbers 8

47530 – dividend   5 – divisor   9506 – quotient

Every time bring down just one digit

Example

```
    9506
5 | 47530
   -45
    25
   -25
     03
     -0
      30
    - 30
       0
```

1)	$5 \times \underline{9} = 45$	Put 9 in quotient
2)	$47 - 45 = 2$	
3)	Bring down the 5	
4)	$5 \times \underline{5} = 25$	Put 5 in quotient
5)	$25 - 25 = 0$	
6)	Bring down the 3	
7)	$5 \times \underline{0} = 0$	Put 0 in quotient
8)	$3 - 0 = 3$	
9)	Bring down the 0	
10)	$5 \times \underline{6} = 30$	Put 6 in quotient
11)	$30 - 30 = 0$	

Review 1 B
Divide Numbers

	a	b	c	d
# 6	55200 ÷ 6	64435 ÷7	74400 ÷8	83745 ÷9
# 7	55206 ÷ 6	64442 ÷7	74408 ÷8	83754 ÷9
# 8	55212 ÷ 6	64449 ÷7	74416 ÷8	83763 ÷9
# 9	55218 ÷ 6	64456 ÷7	74424 ÷8	83772 ÷9
# 10	55224 ÷ 6	64463 ÷7	74432 ÷8	83781 ÷9

Grade 6 Chapter 3 Lesson 8
Division Numbers 8
Every time bring down just one digit
Example

Work	Step	Operation	Note
$\underline{105}$	1)	$15 \times \underline{1} = 15$	Put 1 in quotient
15 | 1575	2)	$15 - 15 = 0$	
$\underline{-15}$	3)	Bring down the 7	
07	4)	$15 \times \underline{0} = 0$	Put 0 in quotient
$\underline{-0}$	5)	$7 - 0 = 7$	
75	6)	Bring down the 5	
$\underline{-75}$	7)	$15 \times \underline{5} = 75$	Put 5 in quotient
0	8)	$75 - 75 = 0$	

Review 2 A
Divide Numbers

	a	b	c	d
# 1	1200÷ 12	1365 ÷ 13	1526÷ 14	1560 ÷ 15
# 2	1212 ÷12	1378 ÷13	1512 ÷14	1545 ÷15
# 3	1224 ÷ 12	1391 ÷ 13	1498 ÷ 14	1530 ÷15
# 4	1236 ÷ 12	1404 ÷ 13	1484 ÷ 14	1515 ÷ 15
# 5	1248 ÷ 12	1417 ÷ 13	1470 ÷ 14	1500 ÷15

Grade 6 Chapter 3 Lesson 8
Division Numbers 8
Every time bring down just one digit
Example

```
    105
18 |1890
   -18
    09
    -0
     90
    -90
      0
```

1) $18 \times \underline{1} = 18$ Put 1 in quotient
2) $18 - 18 = 0$
3) Bring down the 9
4) $18 \times \underline{0} = 0$ Put 0 in quotient
5) $9 - 0 = 9$
6) Bring down the 0
7) $18 \times \underline{5} = 90$ Put 5 in quotient
8) $90 - 90 = 0$

Review 2 B
Divide Numbers

	a	b	c	d
# 6	1680 ÷16	1768 ÷17	1800 ÷18	1995 ÷19
# 7	1696 ÷ 16	1751 ÷17	1818 ÷18	2014 ÷19
# 8	1712 ÷ 16	1734 ÷ 17	1836 ÷18	2033 ÷19
# 9	1728 ÷16	1717 ÷17	1854 ÷18	2052 ÷19
# 10	1744 ÷ 16	1700 ÷ 17	1872 ÷18	2071 ÷19

Grade 6 Chapter 3 Lesson 9
Division Numbers 9
Every time bring down just one digit
Example

```
     205
15 | 3075
    -30
     07
     -0
      75
    - 75
       0
```

1) $15 \times \underline{2} = 30$ Put 2 in quotient
2) $30 - 30 = 0$
3) Bring down the 7
4) $15 \times \underline{0} = 0$ Put 0 in quotient
5) $7 - 0 = 7$
6) Bring down the 5
7) $15 \times \underline{5} = 75$ Put 5 in quotient
8) $75 - 75 = 0$

Review 1 A
Divide Numbers

	a	b	c	d
# 1	2400 ÷ 12	2665 ÷ 13	2926 ÷ 14	3060 ÷ 15
# 2	2412 ÷ 12	2678 ÷ 13	2912 ÷ 14	3045 ÷ 15
# 3	2424 ÷ 12	2691 ÷ 13	2898 ÷ 14	3030 ÷ 15
# 4	2436 ÷ 12	2704 ÷ 13	2884 ÷ 14	3015 ÷ 15
# 5	2448 ÷ 12	2717 ÷ 13	2870 ÷ 14	3000 ÷ 15

Grade 6 Chapter 3 Lesson 9
Division Numbers 9
Every time bring down just one digit
Example

205	1)	$18 \times \underline{2} = 36$	Put 2 in quotient
18 ⟌ 3690	2)	$36 - 36 = 0$	
-36	3)	Bring down the 9	
09	4)	$18 \times \underline{0} = 0$	Put 0 in quotient
-0	5)	$9 - 0 = 9$	
90	6)	Bring down the 0	
-90	7)	$18 \times \underline{5} = 90$	Put 5 in quotient
0	8)	$90 - 90 = 0$	

Review 1 B
Divide Numbers

	a	b	c	d
# 6	3280 ÷16	3468 ÷17	3600 ÷18	3895 ÷19
# 7	3296 ÷ 16	3451 ÷17	3618 ÷18	3914 ÷19
# 8	3312 ÷ 16	3434 ÷ 17	3636 ÷18	3933 ÷19
# 9	3328 ÷16	3417 ÷17	3654 ÷18	3952 ÷19
# 10	3344 ÷ 16	3400 ÷ 17	3672 ÷18	3971 ÷19

Grade 6 Chapter 3 Lesson 9
Division Numbers 9
Every time bring down just one digit
Example

305	1)	$25 \times \underline{3} = 75$	Put 3 in quotient
25 ⟌ 7625	2)	$76 - 75 = 1$	
-75	3)	Bring down the 2	
12	4)	$25 \times \underline{0} = 0$	Put 0 in quotient
-0	5)	$12 - 0 = 12$	
125	6)	Bring down the 5	
-125	7)	$25 \times \underline{5} = 125$	Put 5 in quotient
0	8)	$125 - 125 = 0$	

Review 2 A
Divide Numbers

	a	b	c	d
# 1	6600÷ 22	7015 ÷ 23	7416÷ 24	7600 ÷ 25
# 2	6622 ÷22	7038 ÷23	7392 ÷24	7575 ÷25
# 3	6644 ÷ 22	7061 ÷ 23	7368 ÷ 24	7550 ÷25
# 4	6666 ÷ 22	7084 ÷ 23	7344 ÷ 24	7525 ÷ 25
# 5	6688 ÷ 22	7107 ÷ 23	7320 ÷ 24	7500 ÷25

Grade 6 Chapter 3 Lesson 9
Division Numbers 9
Every time bring down just one digit
Example

```
    305
28 |8540
   -84
    14
    -0
    140
   -140
      0
```

1) $28 \times \underline{3} = 84$ Put 3 in quotient
2) $85 - 84 = 1$
3) Bring down the 4
4) $28 \times \underline{0} = 0$ Put 0 in quotient
5) $14 - 0 = 14$
6) Bring down the 0
7) $28 \times \underline{5} = 140$ Put 5 in quotient
8) $140 - 140 = 0$

Review 2 B
Divide Numbers

	a	b	c	d
# 6	7930 ÷26	8208 ÷27	8400 ÷28	8845 ÷29
# 7	7956 ÷ 26	8181 ÷27	8428 ÷28	8874 ÷29
# 8	7982 ÷ 26	8154 ÷ 27	8456 ÷28	8903 ÷29
# 9	8008 ÷26	8127 ÷27	8484 ÷28	8932 ÷29
# 10	8034 ÷ 26	8100 ÷ 27	8512 ÷28	8961 ÷29

Grade 6 Chapter 3 Lesson 10

Test 3 A

Divide Number

1 $1248 \div 12$

2 $1530 \div 5$

3 $2535 \div 5$

4 $10175 \div 5$

5 $20175 \div 5$

6 $24120 \div 4$

7 $24075 \div 3$

8 $1560 \div 15$

9 $7416 \div 24$

10 $8845 \div 29$

Grade 6 Chapter 3 Lesson 10

Test 3 B

Divide Numbers

1 $1995 \div 19$

2 $1535 \div 5$

3 $2520 \div 5$

4 $8120 \div 4$

5 $16120 \div 4$

6 $18078 \div 3$

7 $40185 \div 5$

8 $1526 \div 14$

9 $8008 \div 26$

10 $8787 \div 29$

Grade 6 Answers Chapter 3

Lesson 1 Review 1 A

	a	b	c	d		a	b	c	d
1)	0	1	9	0	6)	9	8	1	2
2)	2	3	8	7	7)	6	7	3	4
3)	5	4	6	5	8)	4	5	5	6
4)	7	6	4	3	9)	3	2	7	8
5)	8	9	2	1	10)	1	0	0	9

Lesson 1 Review 1 B

	a	b	c	d	e	f	g	h
11)	2	3	4	5	6	7	8	9
12)	4	6	8	10	12	14	16	18
13)	6	9	12	15	18	21	24	27
14)	8	12	16	20	24	28	32	36
15)	10	15	20	25	30	35	40	45
16)	12	18	24	30	36	42	48	54
17)	14	21	28	35	42	49	56	63
18)	16	24	32	40	48	56	64	72
19)	18	27	36	45	54	63	72	81
20)	20	30	40	50	60	70	80	90

Lesson 1 Review 2 A **Review 2 B**

	a	b	c	d		a	b	c	d
1)	100	105	109	104	6)	105	104	100	105
2)	101	106	108	103	7)	106	103	101	106
3)	102	107	107	102	8)	107	102	102	107
4)	103	108	106	101	9)	108	101	103	108
5)	104	109	105	100	10)	109	100	104	109

Lesson 2 Review 1 A | **Review 1 B**

	a	b	c	d		a	b	c	d
1)	200	201	202	203	1)	250	251	252	253
2)	204	205	206	207	2)	254	255	256	257
3)	210	211	212	213	3)	260	261	262	263
4)	214	215	216	217	4)	264	265	266	267
5)	220	221	222	223	5)	270	271	272	273
6)	224	225	226	227	6)	274	275	276	277
7)	230	231	232	233	7)	280	281	282	283
8)	234	235	236	237	8)	284	285	286	287
9)	240	241	242	243	9)	290	291	292	293
10)	244	245	246	247	10)	294	295	296	297

Lesson 2 Review 2 A | **Review 2 B**

	a	b	c	d		a	b	c	d
1)	300	301	302	303	1)	350	351	352	353
2)	304	305	306	307	2)	354	355	356	357
3)	310	311	312	313	3)	360	361	362	363
4)	314	315	316	317	4)	364	365	366	367
5)	320	321	322	323	5)	370	371	372	373
6)	324	325	326	327	6)	374	375	376	377
7)	330	331	332	333	7)	380	381	382	383
8)	334	335	336	337	8)	384	385	386	387
9)	340	341	342	343	9)	390	391	392	393
10)	344	345	346	347	10)	394	395	396	397

Lesson 3 Review 1 A | **Review 1 B**

	a	b	c	d		a	b	c	d
1)	400	401	402	403	1)	450	451	452	453
2)	404	405	406	407	2)	454	455	456	457
3)	410	411	412	413	3)	460	461	462	463
4)	414	415	416	417	4)	464	465	466	467
5)	420	421	422	423	5)	470	471	472	473
6)	424	425	426	427	6)	474	475	476	477
7)	430	431	432	433	7)	480	481	482	483
8)	434	435	436	437	8)	484	485	486	487
9)	440	441	442	443	9)	490	491	492	493
10)	444	445	446	447	10)	494	495	496	497

Grade 6 Answers Chapter 3

Lesson 3 **Review 2 A** **Review 2 B**

	a	b	c	d		a	b	c	d
1)	500	501	502	503	1)	550	551	552	553
2)	504	505	506	507	2)	554	555	556	557
3)	510	511	512	513	3)	560	561	562	563
4)	514	515	516	517	4)	564	565	566	567
5)	520	521	522	523	5)	570	571	572	573
6)	524	525	526	527	6)	574	575	576	577
7)	530	531	532	533	7)	580	581	582	583
8)	534	535	536	537	8)	584	585	586	587
9)	540	541	542	543	9)	590	591	592	593
10)	544	545	546	547	10)	594	595	596	597

Lesson 4

Review 1 A **Review 1 B**

	a	b	c	d		a	b	c	d
1)	1020	1025	1030	1035	6)	1200	1205	1300	1305
2)	1021	1026	1031	1036	7)	1201	1206	1301	1306
3)	1022	1027	1032	1037	8)	1202	1207	1302	1307
4)	1023	1028	1033	1038	9)	1203	1208	1303	1308
5)	1024	1029	1034	1039	10)	1204	1209	1304	1309

Lesson 4 **Review 2 A** **Review 2 B**

	a	b	c	d		a	b	c	d
1)	2020	2025	2030	2035	6)	2200	2205	2300	2305
2)	2021	2026	2031	2036	7)	2201	2206	2301	2306
3)	2022	2027	2032	2037	8)	2202	2207	2302	2307
4)	2023	2028	2033	2038	9)	2203	2208	2303	2308
5)	2024	2029	2034	2039	10)	2204	2209	2304	2309

Grade 6 Answers Chapter 3

Lesson 5 Review 1 A / **Review 1 B**

	a	b	c	d		a	b	c	d
# 1	3020	3025	3030	3035	# 6	3200	3205	3300	3305
# 2	3021	3026	3031	3036	# 7	3201	3206	3301	3306
# 3	3022	3027	3032	3037	# 8	3202	3207	3302	3307
# 4	3023	3028	3033	3038	# 9	3203	3208	3303	3308
# 5	3024	3029	3034	3039	# 10	3204	3209	3304	3309

Lesson 5 Review 2 A / **Review 2 B**

	a	b	c	d		a	b	c	d
# 1	4020	4025	4030	4035	# 6	4200	4205	4300	4305
# 2	4021	4026	4031	4036	# 7	4201	4206	4301	4306
# 3	4022	4027	4032	4037	# 8	4202	4207	4302	4307
# 4	4023	4028	4033	4038	# 9	4203	4208	4303	4308
# 5	4024	4029	4034	4039	# 10	4204	4209	4304	4309

Lesson 6 Review 1 A / **Review 1 B**

	a	b	c	d		a	b	c	d
# 1	5020	5025	5030	5035	# 6	5200	5205	5300	5305
# 2	5021	5026	5031	5036	# 7	5201	5206	5301	5306
# 3	5022	5027	5032	5037	# 8	5202	5207	5302	5307
# 4	5023	5028	5033	5038	# 9	5203	5208	5303	5308
# 5	5024	5029	5034	5039	# 10	5204	5209	5304	5309

Lesson 6 Review 2 A / **Review 2 B**

	a	b	c	d		a	b	c	d
# 1	6020	6025	6030	6035	# 6	6200	6205	6300	6305
# 2	6021	6026	6031	6036	# 7	6201	6206	6301	6306
# 3	6022	6027	6032	6037	# 8	6202	6207	6302	6307
# 4	6023	6028	6033	6038	# 9	6203	6208	6303	6308
# 5	6024	6029	6034	6039	# 10	6204	6209	6304	6309

Grade 6 Answers Chapter 3

Lesson 7 **Review 1 A** **Review 1 B**

	a	b	c	d		a	b	c	d
# 1	7020	7025	7030	7035	# 6	7200	7205	7300	7305
# 2	7021	7026	7031	7036	# 7	7201	7206	7301	7306
# 3	7022	7027	7032	7037	# 8	7202	7207	7302	7307
# 4	7023	7028	7033	7038	# 9	7203	7208	7303	7308
# 5	7024	7029	7034	7039	# 10	7204	7209	7304	7309

Lesson 7 **Review 2 A** **Review 2 B**

	a	b	c	d		a	b	c	d
# 1	8020	8025	8030	8035	# 6	8200	8205	8300	8305
# 2	8021	8026	8031	8036	# 7	8201	8206	8301	8306
# 3	8022	8027	8032	8037	# 8	8202	8207	8302	8307
# 4	8023	8028	8033	8038	# 9	8203	8208	8303	8308
# 5	8024	8029	8034	8039	# 10	8204	8209	8304	8309

Lesson 8 **Review 1 A** **Review 1 B**

	a	b	c	d		a	b	c	d
# 1	9020	9025	9030	9035	# 6	9200	9205	9300	9305
# 2	9021	9026	9031	9036	# 7	9201	9206	9301	9306
# 3	9022	9027	9032	9037	# 8	9202	9207	9302	9307
# 4	9023	9028	9033	9038	# 9	9203	9208	9303	9308
# 5	9024	9029	9034	9039	# 10	9204	9209	9304	9309

Lesson 8 **Review 2 A** **Review2 B**

	a	b	c	d		a	b	c	d
# 1	100	105	109	104	# 6	105	104	100	105
# 2	101	106	108	103	# 7	106	103	101	106
# 3	102	107	107	102	# 8	107	102	102	107
# 4	103	108	106	101	# 9	108	101	103	108
# 5	104	109	105	100	# 10	109	100	104	109

Grade 6 Answers Chapter 3

Lesson 9 Review 1 A

	a	b	c	d
# 1	200	205	209	204
# 2	201	206	208	203
# 3	202	207	207	202
# 4	203	208	206	201
# 5	204	209	205	200

Review 1 B

	a	b	c	d
# 6	205	204	200	205
# 7	206	203	201	206
# 8	207	202	202	207
# 9	208	201	203	208
# 10	209	200	204	209

Lesson 9 Review 2 A

	a	b	c	d
# 1	300	305	309	304
# 2	301	306	308	303
# 3	302	307	307	302
# 4	303	308	306	301
# 5	304	309	305	300

Review 2 B

	a	b	c	d
# 6	305	304	300	305
# 7	306	303	301	306
# 8	307	302	302	307
# 9	308	301	303	308
# 10	309	300	304	309

Lesson 10 Test 3 A

1 104 # 2 306 # 3 507 # 4 2035 # 5 4035
6 6030 # 7 8025 # 8 104 # 9 309 # 10 305

Test 3 B

1 105 # 2 307 # 3 504 # 4 2030 # 5 4030
6 6026 # 7 8037 # 8 109 # 9 308 # 10 303

Grade 6

Chapter 4 Four Operations With Numbers

Grade 6 Chapter 4 Contents
Four Operations with Numbers
Lesson 1 – 10
Basic Problems

Work with lesson 1, read rules and examples, 2) Solve the problems from contents, 3) Check answers in the end of chapter 1, 4) Work with the next lessons 2, 3, 4, 5, …, 5) Solve Test 1 A, 6) Grading Scale: A =10 points, B=9points, C=8 points, D=7 points, F=0-6 points, 7) If you solve 10 problems correctly, you get an A, 8) If have an A, you can move on to next chapter, 9) If your grade is lower than an A, work with the mistakes, then solve Test 1B, 10) When you finish all of the chapters, solve Final Exam in last chapter.

Grade 6 Chapter 4 Lesson 1
Addition Numbers
To add numbers, add in any order
Examples 1

7000 + 35000 + 3000 = 7000 + 38000 = 45000

7000 + 35000 + 3000 = 42000 + 3000 = 45000

7000 + 35000 + 3000=7000 + 3000 + 35000=10000 + 35000=45000

Add in Column 2
Example

$$\begin{array}{r} 35000 \\ +\ 7000 \\ \underline{3000} \\ 45000 \end{array}$$

Review 1 A
Add Numbers

	a	b
# 1	1000 + 9000 + 10000	10000 + 9000 + 1000
# 2	2000 + 8000 + 15000	15000 + 8000 + 2000
# 3	3000 + 7000 + 20000	20000 + 7000 + 3000
# 4	4000 + 6000 + 25000	25000 + 6000 + 4000
# 5	5000 + 5000 + 30000	30000 + 5000 + 5000
# 6	9000 + 1000 + 35000	35000 + 1000 + 9000
# 7	8000 + 2000 + 40000	40000 + 2000 + 8000
# 8	7000 + 3000 + 45000	45000 + 3000 + 7000
# 9	6000 + 4000 + 50000	50000 + 4000 + 6000
# 10	5000 + 5000 + 55000	55000 + 5000 + 5000

Add Numbers

	a	b
# 11	1000 + 60000 + 9000	1000 + 80000 + 9000
# 12	2000 + 65000 + 8000	2000 + 75000 + 8000
# 13	3000 + 70000 + 7000	3000 + 70000 + 7000
# 14	4000 + 75000 + 6000	4000 + 65000 + 6000
# 15	5000 + 80000 + 5000	5000 + 60000 + 5000
# 16	9000 + 60000 + 1000	9000 + 55000 + 1000
# 17	8000 + 65000 + 2000	8000 + 50000 + 2000
# 18	7000 + 70000 + 3000	7000 + 45000 + 3000
# 19	6000 + 75000 + 4000	6000 + 40000 + 4000
# 20	5000 + 80000 + 5000	5000 + 35000 + 5000

Grade 6 Chapter 4 Lesson 1
Addition and Subtraction numbers
To add and subtract numbers work from left to right

Example 1

7000 + 35000 - 3000 = 42000 - 3000 = 39000

Example 2

7000 + 35000 - 3000 = 39000

1)	2)
35000	42000
+ 7000	-3000
42000	39000

Review 1 B

Add and Subtract

	a	b
# 1	20000 + 1000 - 9000	10000 + 9000 - 1000
# 2	25000 + 1000 - 5000	15000 + 8000 - 2000
# 3	30000 + 3000 - 7000	20000 + 7000 - 3000
# 4	35000 + 4000 - 6000	25000 + 6000 - 4000
# 5	40000 + 5000 - 5000	30000 + 5000 - 5000
# 6	45000 + 9000 - 1000	35000 + 1000 - 9000
# 7	50000 + 8000 - 2000	40000 + 2000 - 8000
# 8	55000 + 7000 - 3000	45000 + 3000 - 7000
# 9	60000 + 6000 - 4000	50000 + 4000 - 6000
# 10	65000 + 5000 - 5000	55000 + 5000 - 5000

Add and Subtract

	a	b
# 11	1000 + 50000 - 9000	1000 + 80000 - 9000
# 12	2000 + 55000 - 8000	2000 + 75000 - 8000
# 13	3000 + 60000 - 7000	3000 + 70000 - 7000
# 14	4000 + 65000 - 6000	4000 + 65000 - 6000
# 15	5000 + 70000 - 5000	5000 + 60000 - 5000
# 16	9000 + 60000 - 1000	9000 + 55000 - 1000
# 17	8000 + 65000 - 2000	8000 + 50000 - 2000
# 18	7000 + 70000 - 3000	7000 + 45000 - 3000
# 19	6000 + 75000 - 4000	6000 + 40000 - 4000
# 20	5000 + 80000 - 5000	5000 + 35000 - 5000

Grade 6 Chapter 4 Lesson 2
Subtraction and Addition numbers
To subtract and add numbers work from left to right

Example 1

35000 – 3000 + 7000 = 32000 + 7000 = 39000

Example 2

35000 – 3000 + 7000 = 32000 + 7000 = 39000

1)	2)
35000	32000
- 3000	+7000
32000	39000

Review 1 A
Subtract and Add

	a	b
# 1	20000 – 9000 + 1000	10000 – 1000 + 9000
# 2	25000 – 5000 + 1000	15000 – 2000 + 8000
# 3	30000 – 7000 + 3000	20000 – 3000 + 7000
# 4	35000 - 6000 + 4000	25000 – 4000 + 6000
# 5	40000 - 5000 + 5000	30000 - 5000 + 5000
# 6	45000 – 1000 + 9000	35000 – 9000 + 1000
# 7	50000 – 2000 + 8000	40000 – 8000 + 2000
# 8	55000 - 3000+ 7000	45000 – 7000 + 3000
# 9	60000 - 4000 + 6000	50000 – 6000 + 4000
# 10	65000 - 5000 + 5000	55000 - 5000 + 5000

Subtract and Add

	a	b
# 11	35000 – 9000 + 1000	45000 – 9000 + 1000
# 12	40000 – 8000 + 2000	50000 – 8000 + 2000
# 13	45000 – 7000 + 3000	55000 – 7000 + 3000
# 14	50000 – 6000 + 4000	60000 – 6000 + 4000
# 15	55000 – 5000 + 5000	65000 – 5000 + 5000
# 16	60000 – 1000 + 9000	70000 – 1000 + 9000
# 17	65000 – 2000 + 8000	75000 – 2000 + 8000
# 18	70000 – 3000 + 7000	80000 – 3000 + 7000
# 19	75000 – 4000 + 6000	85000 – 4000 + 6000
# 20	80000 – 5000 + 5000	90000 – 5000 + 5000

Grade 6 Chapter 4 Lesson 2
Subtraction numbers
To subtract numbers, work from left to right
Example 1
35000 – 3000 - 7000 = 32000 - 7000 = 25000
Subtraction numbers in column
Example 2
35000 – 3000 - 7000 = 32000 - 7000 = 25000

1)
$$\begin{array}{r} 35000 \\ -\ 3000 \\ \hline 32000 \end{array}$$

2)
$$\begin{array}{r} 32000 \\ -\ 7000 \\ \hline 25000 \end{array}$$

Review 1 B
Subtract numbers

	a	b
# 1	20000 – 9000 - 1000	10000 – 1000 - 9000
# 2	25000 – 5000 - 1000	15000 – 2000 - 8000
# 3	30000 – 7000 - 3000	20000 – 3000 - 7000
# 4	35000 - 6000 - 4000	25000 – 4000 - 6000
# 5	40000 - 5000 - 5000	30000 - 5000 - 5000
# 6	45000 – 1000 - 9000	35000 – 9000 - 1000
# 7	50000 – 2000 - 8000	40000 – 8000 - 2000
# 8	55000 - 3000 - 7000	45000 – 7000 - 3000
# 9	60000 - 4000 - 6000	50000 – 6000 - 4000
# 10	65000 - 5000 - 5000	55000 - 5000 - 5000

Subtract numbers

	a	b
# 11	35000 – 9000 - 1000	45000 – 9000 - 1000
# 12	40000 – 8000 - 2000	50000 – 8000 - 2000
# 13	45000 – 7000 - 3000	55000 – 7000 - 3000
# 14	50000 – 6000 - 4000	60000 – 6000 - 4000
# 15	55000 – 5000 - 5000	65000 – 5000 - 5000
# 16	60000 – 1000 - 9000	70000 – 1000 - 9000
# 17	65000 – 2000 - 8000	75000 – 2000 - 8000
# 18	70000 – 3000 - 7000	80000 – 3000 - 7000
# 19	75000 – 4000 - 6000	85000 – 4000 - 6000
# 20	80000 – 5000 - 5000	90000 – 5000 - 5000

Grade 6 Chapter 4 Lesson 3

Multiplication numbers

To multiply numbers, work in any order

Examples

$405 \times 8 \times 9 = 3240 \times 9 = 29160$

$405 \times 8 \times 9 = 405 \times 72 = 29160$

$506 \times 9 \times 8 = 4554 \times 8 = 36432$

$506 \times 9 \times 8 = 506 \times 72 = 36432$

Review 1 A

Multiply Numbers

	a	b
# 1	$502 \times 4 \times 5$	$503 \times 5 \times 5$
# 2	$506 \times 4 \times 5$	$507 \times 5 \times 5$
# 3	$602 \times 4 \times 6$	$603 \times 5 \times 6$
# 4	$606 \times 4 \times 6$	$607 \times 5 \times 6$
# 5	$702 \times 4 \times 7$	$703 \times 5 \times 7$
# 6	$706 \times 4 \times 7$	$707 \times 5 \times 7$
# 7	$802 \times 4 \times 8$	$803 \times 5 \times 8$
# 8	$806 \times 4 \times 8$	$807 \times 5 \times 8$
# 9	$902 \times 4 \times 9$	$903 \times 5 \times 9$
# 10	$906 \times 4 \times 9$	$907 \times 5 \times 9$

Multiply Numbers

	a	b
# 11	$504 \times 6 \times 5$	$505 \times 7 \times 5$
# 12	$508 \times 6 \times 5$	$509 \times 7 \times 5$
# 13	$604 \times 6 \times 6$	$605 \times 7 \times 6$
# 14	$608 \times 6 \times 6$	$609 \times 7 \times 6$
# 15	$704 \times 6 \times 7$	$705 \times 7 \times 7$
# 16	$708 \times 6 \times 7$	$709 \times 7 \times 7$
# 17	$804 \times 6 \times 8$	$805 \times 7 \times 8$
# 18	$808 \times 6 \times 8$	$809 \times 7 \times 8$
# 19	$904 \times 6 \times 9$	$905 \times 7 \times 9$
# 20	$908 \times 6 \times 9$	$909 \times 7 \times 9$

Grade 6 Chapter 4 Lesson 3
Multiplication and Division numbers
To multiply and divide numbers work from left to right

Examples

$309 \times 81 \div 9 = 25029 \div 9 = 2781$

$409 \times 81 \div 9 = 33129 \div 9 = 3681$

Review 1 B
Multiply and Divide numbers

	a	b
# 1	$502 \times 81 \div 9$	$503 \times 72 \div 8$
# 2	$506 \times 81 \div 9$	$507 \times 72 \div 8$
# 3	$602 \times 72 \div 9$	$603 \times 64 \div 8$
# 4	$606 \times 72 \div 9$	$607 \times 64 \div 8$
# 5	$702 \times 63 \div 9$	$703 \times 56 \div 8$
# 6	$706 \times 63 \div 9$	$707 \times 56 \div 8$
# 7	$802 \times 54 \div 9$	$803 \times 48 \div 8$
# 8	$806 \times 54 \div 9$	$807 \times 48 \div 8$
# 9	$902 \times 45 \div 9$	$903 \times 40 \div 8$
# 10	$906 \times 45 \div 9$	$907 \times 40 \div 8$

Multiply and Divide numbers

	a	b
# 11	$504 \times 63 \div 7$	$505 \times 54 \div 6$
# 12	$508 \times 63 \div 7$	$509 \times 54 \div 6$
# 13	$604 \times 56 \div 7$	$605 \times 48 \div 6$
# 14	$608 \times 56 \div 7$	$609 \times 48 \div 6$
# 15	$704 \times 49 \div 7$	$705 \times 42 \div 6$
# 16	$708 \times 49 \div 7$	$709 \times 42 \div 6$
# 17	$804 \times 42 \div 7$	$805 \times 36 \div 6$
# 18	$808 \times 42 \div 7$	$809 \times 36 \div 6$
# 19	$904 \times 35 \div 7$	$905 \times 30 \div 6$
# 20	$908 \times 35 \div 7$	$909 \times 30 \div 6$

Grade 6 Chapter 4 Lesson 4

Division and Multiplication numbers

To divide and multiply numbers work from left to right

Examples

$81 \div 9 \times 309 = 9 \times 309 = 2781$

$81 \div 9 \times 409 = 9 \times 409 = 3681$

Review 1 A

Divide and multiply numbers

	a	b
# 1	$81 \div 9 \times 502$	$72 \div 8 \times 503$
# 2	$81 \div 9 \times 506$	$72 \div 8 \times 507$
# 3	$72 \div 9 \times 602$	$64 \div 8 \times 603$
# 4	$72 \div 9 \times 606$	$64 \div 8 \times 607$
# 5	$63 \div 9 \times 702$	$56 \div 8 \times 703$
# 6	$63 \div 9 \times 706$	$56 \div 8 \times 707$
# 7	$54 \div 9 \times 802$	$48 \div 8 \times 803$
# 8	$54 \div 9 \times 806$	$48 \div 8 \times 807$
# 9	$45 \div 9 \times 902$	$40 \div 8 \times 903$
# 10	$45 \div 9 \times 906$	$40 \div 8 \times 907$

Multiply and Divide numbers

	a	b
# 11	$63 \div 7 \times 504$	$54 \div 6 \times 505$
# 12	$63 \div 7 \times 508$	$54 \div 6 \times 509$
# 13	$56 \div 7 \times 604$	$48 \div 6 \times 605$
# 14	$56 \div 7 \times 608$	$48 \div 6 \times 609$
# 15	$49 \div 7 \times 704$	$42 \div 6 \times 705$
# 16	$49 \div 7 \times 708$	$42 \div 6 \times 709$
# 17	$42 \div 7 \times 804$	$36 \div 6 \times 805$
# 18	$42 \div 7 \times 808$	$36 \div 6 \times 809$
# 19	$35 \div 7 \times 904$	$30 \div 6 \times 905$
# 20	$35 \div 7 \times 908$	$30 \div 6 \times 909$

Grade 6 Chapter 4 Lesson 4
Division numbers
To divide numbers, work from left to right
Examples

64000 ÷ 80 ÷ 8 = 800 ÷ 8 = 100

81000 ÷ 90 ÷ 9 = 900 ÷ 9 = 100

Review 1 B
Divide numbers

	a	b
# 1	400 ÷ 2 ÷ 2	1600 ÷ 4 ÷ 4
# 2	1600 ÷ 4 ÷ 2	6400 ÷ 8 ÷ 4
# 3	3600 ÷ 6 ÷ 2	14400 ÷ 12 ÷ 4
# 4	6400 ÷ 8 ÷ 2	25600 ÷ 16 ÷ 4
# 5	10000 ÷ 10 ÷ 2	40000 ÷ 20 ÷ 4
# 6	14400 ÷ 12 ÷ 2	57600 ÷ 24 ÷ 4
# 7	19600 ÷ 14 ÷ 2	78400 ÷ 28 ÷ 4
# 8	25600 ÷ 16 ÷ 2	10240 ÷ 32÷ 4
# 9	32400 ÷ 18 ÷ 2	12960 ÷ 36 ÷ 4
# 10	40000 ÷ 20 ÷ 2	16000 ÷ 40 ÷ 4

Divide numbers

	a	b
# 11	900 ÷ 3 ÷ 3	2500÷ 5 ÷ 5
# 12	3600 ÷ 6 ÷3	10000 ÷10÷ 5
# 13	8100 ÷ 9 ÷3	22500 ÷15 ÷5
# 14	14400 ÷12÷ 3	40000 ÷20 ÷5
# 15	22500 ÷15÷ 3	62500 ÷25 ÷5
# 16	32400 ÷18÷3	90000 ÷30 ÷5
# 17	44100 ÷21÷3	12250÷ 35 ÷5
# 18	57600 ÷24÷3	16000 ÷40 ÷5
# 19	72900 ÷27÷3	20250 ÷45 ÷5
# 20	90000 ÷30÷3	25000 ÷50 ÷5

Grade 6 Chapter 4 Lesson 5
Multiplication and Addition numbers
To multiply and add numbers always first multiply then add

Example 1

$5006 \times 5 + 4970 = 30000$

1) $5006 \times 5 = 25030$

$$\begin{array}{r} 5006 \\ \times\ \underline{\quad 5} \\ 25030 \end{array}$$

2) $25030 + 4970 = 30000$

$$\begin{array}{r} 25030 \\ \underline{+\ 4970} \\ 30000 \end{array}$$

Example 2

$1960 + 6005 \times 8 = 50000$

1) $6005 \times 8 = 48040$

$$\begin{array}{r} 6005 \\ \times\ \underline{\quad 8} \\ 48040 \end{array}$$

2) $48040 + 1960 = 50000$

$$\begin{array}{r} 48040 \\ \underline{+1960} \\ 50000 \end{array}$$

Review 1 A
Multiply and Add Numbers

	a	b
# 1	$1000 \times 5 + 5000$	$1001 \times 5 + 4995$
# 2	$2002 \times 6 + 7988$	$2003 \times 6 + 7982$
# 3	$3004 \times 7 + 8972$	$3005 \times 7 + 8965$
# 4	$4006 \times 8 + 7952$	$4007 \times 8 + 7944$
# 5	$5008 \times 9 + 4928$	$6009 \times 9 + 5919$

Multiply and Add Numbers

	a	b
# 6	$5000 + 5000 \times 9$	$4991 + 5001 \times 9$
# 7	$11984 + 6002 \times 8$	$11976 + 6003 \times 8$
# 8	$20972 + 7004 \times 7$	$20965 + 7005 \times 7$
# 9	$31964 + 8006 \times 6$	$15944 + 8007 \times 8$
# 10	$44960 + 9008 \times 5$	$44955 + 9009 \times 5$

Grade 6 Chapter 4 Lesson 5
Multiplication and Subtraction numbers
To multiply and subtract numbers always first multiply then subtract

Example 1 $506 \times 80 - 30480 = 10000$

1) $506 \times 80 = 40480$

2) $40480 - 30480 = 10000$

Example 2 $55600 - 608 \times 75 = 10000$

1) 608×75: $3040 + 42560 = 45600$

2) $55600 - 45600 = 10000$

Review 1 B
Multiply and Subtract Numbers

	a	b
# 1	100 × 10 - 0	101 × 15 - 515
# 2	202 × 20 - 3040	203 × 25 - 1075
# 3	304 × 30 – 1120	305 × 35 - 675
# 4	406 × 40 – 6240	407 × 45 - 8315
# 5	508 × 50 – 5400	509 × 55 - 7995
# 6	35000 - 500 × 50	37555 - 501 × 55
# 7	56120 - 602 × 60	59195 - 603 × 65
# 8	79280 - 704 × 70	82875 - 705 × 75
# 9	84480 - 806 × 80	88595 - 807 × 85
# 10	91720 - 908 × 90	96355 - 909 × 95

Grade 6 Chapter 4 Lesson 6
Division and Addition numbers
To divide and add numbers always first divide then add

Example 1

2 1
696 + 1520 ÷ 5 = 1000

```
      304
1) 5)1520    2)  696
    -15         +304
      20        1000
     -20
```

Example 2

1 2
10020 ÷ 5 + 7996 = 10000

```
1)  2004       2) 7996
  5)10020       + 2004
   -10          10000
     020
     -20
```

Review 1 A
Divide and Add Numbers

	a	b
# 1	9207 ÷ 9 + 8977	9288 ÷ 9 + 8968
# 2	16272 ÷ 8 + 17966	16344 ÷ 8 + 17957
# 3	21315 ÷ 7 + 26955	21378 ÷ 7 + 26946
# 4	24336 ÷ 6 + 15944	24390 ÷ 6 + 15935
# 5	25335 ÷ 5 + 4933	25380 ÷ 6 + 5770

Divide and Add Numbers

	a	b
# 6	4393 + 28035 ÷ 5	4294 + 28530 ÷ 5
# 7	15494 + 27036 ÷ 6	15395 + 27630 ÷ 6
# 8	26595 + 23835÷ 7	26496 + 24528 ÷ 7
# 9	17696 + 18432÷ 8	17597 + 19224 ÷ 8
# 10	8797 + 10827÷ 9	8698 + 11718 ÷ 9

Grade 6 Chapter 4 Lesson 6
Division and Subtraction numbers
To divide and subtract numbers always first divide after that subtract
Always do multiplication or division before addition or subtraction

Example 1

1 2

24000 ÷ 80 - 200 = 100

1) **24000 ÷ 80 = 2400 ÷ 8 = 300**
2) **300 – 200 = 100**

Example 2

2 1

720 - 1600 ÷ 80 = 720 – 20 = 700

1) **1600 ÷ 80 = 160 ÷ 8 = 20**
2) **720 – 20 = 700**

Review 1 B

Evaluate

	a	b
# 1	50000 ÷ 100 - 50	10000 ÷ 50 - 100
# 2	60000 ÷ 200 – 50	18000 ÷ 60 - 100
# 3	60000 ÷ 300 - 50	35000 ÷ 70 - 200
# 4	80000 ÷ 400 - 100	40000 ÷ 80 – 300
# 5	80000 ÷ 500 – 60	36000 ÷ 90 - 300

Evaluate

	a	b
# 6	10000 - 5000 ÷ 10	10000 - 5000 ÷ 50
# 7	12000 - 6000 ÷ 20	12000 - 6000 ÷ 60
# 8	18000 - 6000 ÷ 30	14000 - 7000 ÷ 70
# 9	18000 - 8000 ÷ 40	16000 - 8000 ÷ 80
# 10	19000 - 9000 ÷ 50	18000 - 9000 ÷ 90

Grade 6 Chapter 4 Lesson 7
Addition and Multiplication with Parentheses
Work always in parentheses first

Example 1

$(17 + 19) + 34 = 36 + 34 = 70$

$43 + (18 + 19) = 43 + 37 = 80$

Example 2

$(25 + 5) \times 8 = 30 \times 8 = 240$

$6 \times (125 + 25) = 6 \times 150 = 900$

Review 1 A

Evaluate

	a	b
# 1	(1 + 9) + 10	18 + (13 + 19)
# 2	(2 + 8) + 10	18 + (14 + 18)
# 3	(3 + 7) + 10	18 + (15 + 17)
# 4	(4 + 6) + 10	15 + (16 + 19)
# 5	(5 + 5) + 10	15 + (17 + 18)
# 6	(2 + 9) + 19	27 + (14 + 19)
# 7	(3 + 8) + 19	27 + (15 + 18)
# 8	(4 + 7) + 19	27 + (16 + 17)
# 9	(5 + 6) + 19	26 + (15 + 19)
# 10	(6 + 9) + 15	26 + (16 + 18)

Evaluate

	a	b
# 11	(20 × 9) × 5	5 × (10 + 6)
# 12	(50 × 6) × 2	(20 + 6) × 5
# 13	(8 × 4) × 25	5 × (30 + 8)
# 14	(25 × 6) × 4	(40 + 8) × 5
# 15	(25 × 8) × 4	25× (12 + 18)
# 16	25 × (4 × 5)	(10 – 2) × 25
# 17	8 × (4 × 25)	25 × (10 – 4)
# 18	25 × (4 × 5)	(20 – 2) × 25
# 19	5 × (25 × 2)	8 × (20 – 5)
# 20	25 × (5 × 4)	(30 – 5) × 8

Grade 6 Chapter 4 Lesson 7
Addition and Subtraction with Parentheses
Always first, work in parentheses
Examples

(700 + 300) - 300= 1000 - 300 = 700

(700 - 300) + 300 = 400 + 300 = 700

9000 – (500 + 500) = 9000 – 1000 = 8000

(1200 – 500) – 500 = 700 – 500 = 200

Review 1 B
Evaluate

	a	b
# 1	(100 + 900) - 100	1600 + (900 - 500)
# 2	(200 + 800) - 150	2600 + (800 – 400)
# 3	(300 + 700) - 200	3600 + (700 – 300)
# 4	(400 + 600) - 250	4600 + (600 – 200)
# 5	(500 + 500) - 300	5600 + (500 – 100)
# 6	(900 + 100) - 350	2600 + (900 – 500)
# 7	(800 + 200) - 400	3600 + (800 – 400)
# 8	(700 + 300) - 450	4600 + (700 – 300)
# 9	(600 + 400) - 500	5600 + (600 – 200)
# 10	(500 + 500) - 550	6600 + (500 – 100)

Evaluate

	a	b
# 11	6000 – (900+100)	(900 - 500) - 100
# 12	6500 – (800+200)	(800 – 400) - 200
# 13	7000 – (700 + 300)	(700 - 300) - 300
# 14	7500 – (600 + 400)	(600 - 200) - 200
# 15	8000 – (500+ 500)	(500 – 100) - 100
# 16	6000 – (100+900)	(1000 – 500) - 100
# 17	6500 – (200+800)	(900 - 400) - 200
# 18	7000 – (300+700)	(800 – 300) - 300
# 19	7500 – (400+600)	(700 - 200) - 200
# 20	8000 – (500+500)	(600- 100) - 100

Grade 6 Chapter 4 Lesson 8
Multiplication and Division with Parentheses
Always first, work in parentheses

$706 \times (36 \div 4) = 706 \times 9 = 6354$

$24028 \div (16 \div 4) = 24028 \div 4 = 6007$

Review 1 A
Evaluate

	a	b
# 1	$102 \times (81 \div 9)$	$607 \times (56 \div 8)$
# 2	$203 \times (72 \div 8)$	$708 \times (54 \div 9)$
# 3	$304 \times (72 \div 9)$	$809 \times (48 \div 8)$
# 4	$405 \times (64 \div 8)$	$908 \times (45 \div 9)$
# 5	$506 \times (63 \div 9)$	$807 \times (40 \div 8)$

Evaluate

	a	b
# 6	$9018 \div (36 \div 4)$	$14007 \div (28 \div 4)$
# 7	$18027 \div (27 \div 3)$	$18012 \div (24 \div 4)$
# 8	$24032 \div (32 \div 4)$	$24018 \div (30 \div 5)$
# 9	$32040 \div (24 \div 3)$	$25020 \div (25 \div 5)$
# 10	$35042 \div (35 \div 5)$	$20025 \div (20 \div 4)$

Grade 6 Chapter 4 Lesson 8

Expressions with Parentheses

Always first, work in parentheses

Examples

$15 \times (807 - 200) = 15 \times 607 = 9105$

$(9000 + 105) \div 15 = 9105 \div 15 = 607$

$(17472 - 9000) \div 12 = 8472 \div 12 = 706$

$84072 \div (10 + 2) = 84072 \div 12 = 7006$

Review 1 B

Evaluate

	a	b
# 1	$2 \times (202 - 100)$	$(1000 + 1004) \div 2$
# 2	$3 \times (303 - 100)$	$(3000 + 3009) \div 3$
# 3	$4 \times (404 - 100)$	$(10000 + 2016) \div 4$
# 4	$5 \times (505 - 100)$	$(10000 + 10025) \div 5$
# 5	$6 \times (606 - 100)$	$(20000+ 10036) \div 6$
# 6	$7 \times (707 - 100)$	$(30000 + 12049) \div 7$
# 7	$8 \times (808 - 100)$	$(40000 + 16064) \div 8$
# 8	$9 \times (909 - 100)$	$(50000 + 22081) \div 9$
# 9	$10 \times (1008 - 100)$	$(60000 + 30080) \div 10$
# 10	$11 \times (907 - 100)$	$(70000 + 18077) \div 11$

Evaluate

	a	b
# 11	$(304 - 100) \div 2$	$2004 \div (10 - 8)$
# 12	$(809 - 200) \div 3$	$6009 \div (10 - 7)$
# 13	$(2216 - 1000) \div 4$	$12016 \div (10 - 6)$
# 14	$(4025 - 2000) \div 5$	$20025 \div (10 - 5)$
# 15	$(6036 - 3000) \div 6$	$30036 \div (10 - 4)$
# 16	$(8249 - 4000) \div 7$	$42049 \div (10 - 3)$
# 17	$(10664 - 5000) \div 8$	$56064 \div (10 - 2)$
# 18	$(13281 - 6000) \div 9$	$72081 \div (10 - 1)$
# 19	$(16080 - 7000) \div 10$	$90080 \div (10 - 0)$
# 20	$(16877 - 8000) \div 11$	$88077 \div (10 + 1)$

Grade 6 Chapter 4 Lesson 9

Four Operations with numbers

1) Multiplication numbers: for memory <u>(x, x, any)</u>
To multiply numbers, work in any order
$9 \times 8 \times 7 = 72 \times 7 = 504$ $9 \times 8 \times 7 = 9 \times 56 = 504$

2) Work from left to right: for memory <u>(L → R)</u>
a)Multiplication and Division numbers: (× or ÷)
To multiply and divide numbers work from left to right (L → R)
$9 \times 6 \div 3 = 54 \div 3 = 18$ $4 \times 81 \div 9 = 324 \div 9 = 36$
b)Division and Multiplication numbers: (÷ or ×)
To divide and multiply numbers work from left to right <u>(L → R)</u>
$24 \div 6 \times 2 = 4 \times 2 = 8$ $81 \div 9 \times 4 = 9 \times 4 = 36$
c)Division numbers: (÷, ÷)
To divide numbers, work from left to right <u>(L → R)</u>
$24 \div 4 \div 2 = 6 \div 2 = 3$ $810 \div 9 \div 3 = 90 \div 9 = 10$

3) Do Multiplication or Division before Addition or Subtraction
for memory (× or ÷) before (+ or -)
a)Multiplication and Addition numbers: 1) × 2) +
To multiply and add numbers always first multiply than add
$9 \times 6 + 9 = 54 + 9 = 63$ $9 + 9 \times 9 = 9 + 81 = 90$
b)Multiplication and Subtraction numbers: 1) × 2) −
To multiply and subtract numbers always first multiply than subtract
$9 \times 9 - 9 = 81 - 9 = 72$ $90 - 9 \times 9 = 90 - 81 = 9$
c)Division and Addition numbers: 1) ÷ 2) +
To divide and add numbers always first divide than add
$50 \div 5 + 39 = 10 + 39 = 49$ $88 + 171 \div 9 = 88 + 19 = 107$
d)Division and Subtraction numbers: 1) ÷ 2) −
To divide and subtract numbers always first divide than subtract
$108 \div 6 - 9 = 18 - 9 = 9$ $500 - 126 \div 7 = 500 - 18 = 482$

Grade 6 Chapter 4 Lesson 9
Four operations with numbers
Review 4A
Perform the operations

\# 1 $46020 \div 65 - 108$

\# 2 $(25 + 5) \times 8$

\# 3 $64000 \div 80 \div 8$

\# 4 $506 \times 80 - 30480$

\# 5 $10170 \div 5 + 7966$

Review 4 B
Perform the operations

\# 1 $304 \times 81 \div 9$

\# 2 $81 \div 9 \times 305$

\# 3 $706 \times (36 \div 4)$

\# 4 $15 \times (807 - 200)$

\# 5 $(17472 - 9000) \div 12$

Grade 6 Chapter 4 Lesson 10
Four operations with numbers
Test 4 A
Perform the operations

 # 1 $19000 - 9000 \div 50$

 # 2 $6 \times (125 + 25)$

 # 3 $81000 \div 90 \div 9$

 # 4 $55600 - 608 \times 75$

 # 5 $7696 + 11520 \div 5$

Grade 6 Chapter 4 Lesson 10
Four operations with numbers
Test 4 B
Perform the operations

1 $405 \times 81 \div 9$

2 $81 \div 9 \times 406$

3 $24028 \div (16 \div 4)$

4 $(9000 + 105) \div 15$

5 $84072 \div (10 + 2)$

Grade 6 Answers Chapter 4

Lesson 1
Review 1 A

# 1	a)	20000	b)	20000	# 11	a)	70000	b)	90000
# 2	a)	25000	b)	25000	# 12	a)	75000	b)	85000
# 3	a)	30000	b)	30000	# 13	a)	80000	b)	80000
# 4	a)	35000	b)	35000	# 14	a)	85000	b)	75000
# 5	a)	40000	b)	40000	# 15	a)	90000	b)	70000
# 6	a)	45000	b)	45000	# 16	a)	70000	b)	65000
# 7	a)	50000	b)	50000	# 17	a)	75000	b)	60000
# 8	a)	55000	b)	55000	# 18	a)	80000	b)	55000
# 9	a)	60000	b)	60000	# 19	a)	85000	b)	50000
# 10	a)	65000	b)	65000	# 20	a)	90000	b)	45000

Lesson 1
Review 1 B

# 1	a)	12000	b)	18000	# 11	a)	42000	b)	72000
# 2	a)	21000	b)	21000	# 12	a)	49000	b)	69000
# 3	a)	26000	b)	24000	# 13	a)	56000	b)	66000
# 4	a)	33000	b)	27000	# 14	a)	63000	b)	63000
# 5	a)	40000	b)	30000	# 15	a)	70000	b)	60000
# 6	a)	53000	b)	27000	# 16	a)	68000	b)	63000
# 7	a)	56000	b)	34000	# 17	a)	71000	b)	56000
# 8	a)	59000	b)	41000	# 18	a)	74000	b)	49000
# 9	a)	62000	b)	48000	# 19	a)	77000	b)	42000
# 10	a)	65000	b)	55000	# 20	a)	80000	b)	35000

Lesson 2
Review 1 A

# 1	a)	12000	b)	18000	# 11	a)	27000	b)	37000
# 2	a)	21000	b)	21000	# 12	a)	34000	b)	44000
# 3	a)	26000	b)	24000	# 13	a)	41000	b)	51000
# 4	a)	33000	b)	27000	# 14	a)	48000	b)	58000
# 5	a)	40000	b)	30000	# 15	a)	55000	b)	65000
# 6	a)	53000	b)	27000	# 16	a)	68000	b)	78000
# 7	a)	56000	b)	34000	# 17	a)	71000	b)	81000
# 8	a)	59000	b)	41000	# 18	a)	74000	b)	84000
# 9	a)	62000	b)	48000	# 19	a)	77000	b)	87000
# 10	a)	65000	b)	55000	# 20	a)	80000	b)	90000

Grade 6 Answers Chapter 4

Lesson 2 Review 1 B

# 1	a)	10000	b)	0	# 11	a)	25000	b)	35000
# 2	a)	19000	b)	5000	# 12	a)	30000	b)	40000
# 3	a)	20000	b)	10000	# 13	a)	35000	b)	45000
# 4	a)	25000	b)	15000	# 14	a)	40000	b)	50000
# 5	a)	30000	b)	20000	# 15	a)	45000	b)	55000
# 6	a)	35000	b)	25000	# 16	a)	50000	b)	60000
# 7	a)	40000	b)	30000	# 17	a)	55000	b)	65000
# 8	a)	45000	b)	35000	# 18	a)	60000	b)	70000
# 9	a)	50000	b)	40000	# 19	a)	65000	b)	75000
# 10	a)	55000	b)	45000	# 20	a)	70000	b)	80000

Lesson 3 Review 1 A

# 1	a)	10040	b)	12575	# 11	a)	15120	b)	17675
# 2	a)	10120	b)	12675	# 12	a)	15240	b)	17815
# 3	a)	14448	b)	18090	# 13	a)	21744	b)	25410
# 4	a)	14544	b)	18210	# 14	a)	21888	b)	25578
# 5	a)	19656	b)	24605	# 15	a)	29568	b)	34545
# 6	a)	19768	b)	24745	# 16	a)	29736	b)	34741
# 7	a)	25664	b)	32120	# 17	a)	38592	b)	45080
# 8	a)	25792	b)	32280	# 18	a)	38784	b)	45304
# 9	a)	32472	b)	40635	# 19	a)	48816	b)	57015
# 10	a)	32616	b)	40815	# 20	a)	49032	b)	57267

Lesson 3 Review 1 B

# 1	a)	4518	b)	4527	# 11	a)	4536	b)	4545
# 2	a)	4554	b)	4563	# 12	a)	4572	b)	4581
# 3	a)	4816	b)	4824	# 13	a)	4832	b)	4840
# 4	a)	4848	b)	4856	# 14	a)	4864	b)	4872
# 5	a)	4914	b)	4921	# 15	a)	4928	b)	4935
# 6	a)	4942	b)	4949	# 16	a)	4956	b)	4963
# 7	a)	4812	b)	4818	# 17	a)	4824	b)	4830
# 8	a)	4836	b)	4842	# 18	a)	4848	b)	4854
# 9	a)	4510	b)	4515	# 19	a)	4520	b)	4525
# 10	a)	4530	b)	4535	# 20	a)	4540	b)	4545

Grade 6 Answers Chapter 4

Lesson 4 Review 1 A

# 1	a)	4518	b)	4527	# 11	a)	4536	b)	4545
# 2	a)	4554	b)	4563	# 12	a)	4572	b)	4581
# 3	a)	4816	b)	4824	# 13	a)	4832	b)	4840
# 4	a)	4848	b)	4856	# 14	a)	4864	b)	4872
# 5	a)	4914	b)	4921	# 15	a)	4928	b)	4935
# 6	a)	4942	b)	4949	# 16	a)	4956	b)	4963
# 7	a)	4812	b)	4818	# 17	a)	4824	b)	4830
# 8	a)	4836	b)	4842	# 18	a)	4848	b)	4854
# 9	a)	4510	b)	4515	# 19	a)	4520	b)	4525
# 10	a)	4530	b)	4535	# 20	a)	4540	b)	4545

Lesson 4 Review 1 B

# 1	a)	100	b)	100	# 11	a)	100	b)	100
# 2	a)	200	b)	200	# 12	a)	200	b)	200
# 3	a)	300	b)	300	# 13	a)	300	b)	300
# 4	a)	400	b)	400	# 14	a)	400	b)	400
# 5	a)	500	b)	500	# 15	a)	500	b)	500
# 6	a)	600	b)	600	# 16	a)	600	b)	600
# 7	a)	700	b)	700	# 17	a)	700	b)	70
# 8	a)	800	b)	80	# 18	a)	800	b)	80
# 9	a)	900	b)	90	# 19	a)	900	b)	90
# 10	a)	1000	b)	100	# 20	a)	1000	b)	100

Lesson 5 Review 1 A

# 1	a)	10000	b)	10000	# 6	a)	50000	b)	50000
# 2	a)	20000	b)	20000	# 7	a)	60000	b)	60000
# 3	a)	30000	b)	30000	# 8	a)	70000	b)	70000
# 4	a)	40000	b)	40000	# 9	a)	80000	b)	80000
# 5	a)	50000	b)	60000	# 10	a)	90000	b)	90000

Lesson 5
Review 1 B

# 1	a)	1000	b)	1000	# 6	a)	10000	b)	10000
# 2	a)	1000	b)	4000	# 7	a)	20000	b)	20000
# 3	a)	8000	b)	10000	# 8	a)	30000	b)	30000
# 4	a)	10000	b)	10000	# 9	a)	20000	b)	20000
# 5	a)	20000	b)	20000	# 10	a)	10000	b)	10000

Lesson 6
Review 1 A

# 1	a)	10000	b)	10000	# 6	a)	10000	b)	10000
# 2	a)	20000	b)	20000	# 7	a)	20000	b)	20000
# 3	a)	30000	b)	30000	# 8	a)	30000	b)	30000
# 4	a)	20000	b)	20000	# 9	a)	20000	b)	20000
# 5	a)	10000	b)	10000	# 10	a)	10000	b)	10000

Lesson 6
Review 1 B

	a	b		a	b
#1	450	100	#6	9500	9900
#2	250	200	#7	11700	11900
#3	150	300	#8	17800	13900
#4	100	200	#9	17800	15900
#5	100	100	#10	18820	17900

Lesson 7 **Review 1 A**

# 1	a)	20	b)	50	# 11	a)	900	b)	80
# 2	a)	20	b)	50	# 12	a)	600	b)	130
# 3	a)	20	b)	50	# 13	a)	800	b)	190
# 4	a)	20	b)	50	# 14	a)	600	b)	240
# 5	a)	20	b)	50	# 15	a)	800	b)	750
# 6	a)	30	b)	60	# 16	a)	500	b)	200
# 7	a)	30	b)	60	# 17	a)	800	b)	150
# 8	a)	30	b)	60	# 18	a)	500	b)	450
# 9	a)	30	b)	60	# 19	a)	250	b)	120
# 10	a)	30	b)	60	# 20	a)	500	b)	200

Grade 6 Answers Chapter 4

Lesson 7
Review 1 B

# 1	a)	900	b)	2000	# 11	a)	5000	b)	300
# 2	a)	850	b)	3000	# 12	a)	5500	b)	200
# 3	a)	800	b)	4000	# 13	a)	6000	b)	100
# 4	a)	750	b)	5000	# 14	a)	6500	b)	200
# 5	a)	700	b)	6000	# 15	a)	7000	b)	300
# 6	a)	650	b)	3000	# 16	a)	5000	b)	400
# 7	a)	600	b)	4000	# 17	a)	5500	b)	300
# 8	a)	550	b)	5000	# 18	a)	6000	b)	200
# 9	a)	500	b)	6000	# 19	a)	6500	b)	300
# 10	a)	450	b)	7000	# 20	a)	7000	b)	400

Lesson 8
Review 1 A

# 1	a)	918	b)	4249	# 6	a)	1002	b)	2001
# 2	a)	1827	b)	4248	# 7	a)	2003	b)	3002
# 3	a)	2432	b)	4854	# 8	a)	3004	b)	4003
# 4	a)	3240	b)	4540	# 9	a)	4005	b)	5004
# 5	a)	3542	b)	4035	# 10	a)	5006	b)	4005

Lesson 8
Review 1 B

# 1	a)	204	b)	1002	# 11	a)	102	b)	1002
# 2	a)	609	b)	2003	# 12	a)	203	b)	2003
# 3	a)	1216	b)	3004	# 13	a)	304	b)	3004
# 4	a)	2025	b)	4005	# 14	a)	405	b)	4005
# 5	a)	3036	b)	5006	# 15	a)	506	b)	5006
# 6	a)	4249	b)	6007	# 16	a)	607	b)	6007
# 7	a)	5664	b)	7008	# 17	a)	708	b)	7008
# 8	a)	7281	b)	8009	# 18	a)	809	b)	8009
# 9	a)	9080	b)	9008	# 19	a)	908	b)	9008
# 10	a)	8877	b)	8007	# 20	a)	807	b)	8007

Grade 6 Answers Chapter 4

Lesson 9

Review 4 A		Review 4 B	
# 1	600	# 1	2736
# 2	240	# 2	2745
# 3	100	# 3	6354
# 4	10000	# 4	9105
# 5	10000	# 5	706

Lesson 10

Test 4 A		Test 4 B	
# 1	18820	# 1	3645
# 2	900	# 2	3654
# 3	100	# 3	6007
# 4	10000	# 4	607
# 5	10000	# 5	7006

Grade 6

Chapter 5

Finding

Unknowns 6

Grade 6 Chapter 5 Contents
Summary of Basic Skills
Lesson 1 – 16
Advanced Problems

Work with lesson 1, read rules and examples, 2) Solve the problems from contents, 3) Check answers in the end of chapter 1, 4) Work with the next lessons 2, 3, 4, 5, …, 5) Solve Test 1 A, 6) Grading Scale: A =10 points, B=9points, C=8 points, D=7 points, F=0-6 points, 7) If you solve 10 problems correctly, you get an A, 8) If have an A, you can move on to next chapter, 9) If your grade is lower than an A, work with the mistakes, then solve Test 1B, 10) When you finish all of the chapters, solve Final Exam in last chapter.

Grade 6 Chapter 5 Lesson 1
Addition and subtraction
Rules and Examples

Addition

First number in addition called the addend 2 $2 < 5$

Second number in addition called the addend 3 $3 < 5$

Result of addition called the sum 5 $5 > 2$ $5 > 3$

Addend + Addend = Sum $2 + 3 = 5$

First Addend = Sum – Second Addend
$? + 3 = 5$ $? = 5 - 3$ $? = 2$

Second Addend = Sum – First Addend
$2 + ? = 5$ $? = 5 - 2$ $? = 3$

Subtraction

First number in subtraction called the minuend 5 $5 > 2$ $5 > 3$

Second number in subtraction called the subtrahend 2 $2 < 5$

Result of multiplication called the product 3 $3 < 5$

Minuend – Subtrahend = Difference $5 - 2 = 3$

Minuend = Difference + Subtrahend
$? - 2 = 3$ $? = 3 + 2$ $? = 5$

Subtrahend = Minuend – Difference
$5 - ? = 3$ $? = 5 - 3$ $? = 2$

Grade 6 Chapter 5 Lesson 1
Multiplication and division
Rules and Examples

Multiplication

First number in multiplication called the factor **2** $2 < 6$

Second number in multiplication called the factor **3** $3 < 6$

Result of multiplication called the product **6** $6 > 2$ $6 > 3$

Factor × Factor = Product $2 \times 3 = 6$

First Factor = Product ÷ Second Factor
$? \times 3 = 6$ $? = 6 \div 3$ $? = 2$

Second Factor = Product ÷ First Factor
$2 \times ? = 6$ $? = 6 \div 2$ $? = 3$

Division

First number in division is called dividend **6** $6 > 2$ $6 > 3$

Second number in division is called divisor **2** $2 < 6$

Result of division is called quotient **3** $3 < 6$

Dividend ÷ Divisor = Quotient $6 \div 2 = 3$

Divisor = Dividend ÷ Quotient
$6 \div ? = 3$ $? = 6 \div 3$ $? = 2$

Dividend = Quotient × Divisor
$? \div 2 = 3$ $? = 3 \times 2$ $? = 6$

Grade 6 Chapter 5 Lesson 2
Addition and subtraction 1
Rules for addition

Addend + Addend = Sum **2 + 3 = 5 5 > 2 5 > 3**

First Addend = Sum – Second Addend

? + 3 = 5 ? = 5 – 3 ? = 2

Second Addend = Sum – First Addend

2 + ? = 5 ? = 5 – 2 ? = 3

Review 1 A
Put right number instead question mark

	a	b
# 1	? + 1191 = 3110	3939 + ? = 5332
# 2	? + 1282 = 4110	4848 + ? = 6332
# 3	? + 1373 = 5110	5757 + ? = 7332
# 4	? + 1464 = 6110	6969 + ? = 8665
# 5	? + 1595 = 7554	7979 + ? = 9776
# 6	? + 1292 = 4221	4949 + ? = 6443
# 7	? + 1383 = 5221	5858+ ? = 7443
# 8	? + 1474 = 6221	6767+ ? = 8443
# 9	? + 1565 = 7221	7878+ ? = 9665
# 10	? + 1686 = 8554	8989+ ? = 9887

Rules for subtraction

Minuend – Subtrahend = Difference **5 – 2 = 3 5 > 2 5 > 3**

Minuend = Subtrahend + Difference

? - 2 = 3 ? = 3 + 2 ? = 5

Subtrahend = Minuend – Difference

5 - ? = 3 ? = 5 – 3 ? = 2

Put right number instead question mark

	a	b
# 11	? – 1919 =1191	5332 – ? = 1393
# 12	? – 2828=1282	6332– ? = 1484
# 13	? – 3737=1373	7332– ? = 1575
# 14	? – 4646=1464	8665– ? = 1696
# 15	? - 5959=1595	9776- ? = 1797
# 16	? – 2929=1292	6443 –? = 1494
# 17	? – 3838=1383	7443– ? = 1585
# 18	? – 4747=1474	8443– ? = 1676
# 19	? – 5656=1565	9665– ? =1787
# 20	? - 6868=1686	9887- ? = 898

Grade 6 Chapter 5 Lesson 2
Addition and subtraction 1
Rules for addition

Addend + Addend = Sum	2 + 3 = 5	2 < 5	2 < 3
First Addend = Sum – Second Addend	? + 3 = 5	? = 5 – 3	? = 2
Second Addend = Sum – First Addend	2 + ? = 5	? = 5 – 2	? = 3

Review 1 B
Put right number instead question mark

	a	b
# 1	? + 26960 = 76650	46960 + ? = 76650
# 2	? + 27870 = 76650	47870 + ? = 76650
# 3	? + 27970 = 77760	47970 + ? = 77760
# 4	? + 28880 = 77760	45550 + ? = 71100
# 5	? + 29890 = 78870	49990 + ? = 79980
# 6	? + 28980 = 78870	45950 + ? = 75540
# 7	? + 28580 = 74430	45750 + ? = 73320
# 8	? + 29490 = 74430	44840 + ? = 73320
# 9	? + 29390 = 73320	45650 + ? = 72210
# 10	? + 28480 = 73320	44740 + ? = 72210

Rules for subtraction

Minuend – Subtrahend = Difference 5 – 2 = 3 5 > 2 5 > 3

Minuend = Difference + Subtrahend

? - 2 = 3 ? = 3 + 2 ? = 5

Subtrahend = Minuend – Difference

5 - ? = 3 ? = 5 – 3 ? = 2

Put right number instead question mark

	a	b
# 11	? – 29690 = 46960	76650 – ? = 47870
# 12	? – 26960 = 49690	76650 – ? = 48780
# 13	? – 29790 = 47970	77760 – ? = 49790
# 14	? – 28880 = 48880	71100 – ? = 45550
# 15	? – 28980 = 49890	79980 - ? = 49990
# 16	? –29890 = 48980	75540 –? = 49590
# 17	? – 25850 = 48580	73320 – ? = 47570
# 18	? – 24940 = 49490	73320 – ? = 48480
# 19	? – 23930 = 49390	72210 – ? =46560
# 20	? - 24840 = 48480	72210 - ? = 47470

Grade 6 Chapter 5 Lesson 3
Addition and subtraction 2
Rules for addition

Addend + Addend = Sum   2 + 3 = 5   5 > 2   5 > 3

First Addend = Sum – Second Addend

? + 3 = 5   ? = 5 – 3   ? = 2

Second Addend = Sum – First Addend

2 + ? = 5   ? = 5 – 2   ? = 3

Review 1 A
Complete the tables

 # 1

	a	b	c	d	e
Addend		68280	67370		69590
Addend	11910		13730	14640	
Sum	81100	81100		81100	85540

 # 2

Addend		68380	67470		68680
Addend	12920		14740	15650	
Sum	82210	82210		82210	85540

 # 3

	a	b	c	d	e
Addend	?	68480	67570		66860
Addend	13930		15750	16660	
Sum	83320	83320		83320	85540

a) ? + 13930 = 83320

 # 4

Addend	?	68580	66760	?	67770
Addend	14940	?	17670	16760	?
Sum	84430	84430	?	84430	85540

Grade 6 Chapter 5 Lesson 3

Addition and subtraction 2

Rules for subtraction

Minuend – Subtrahend = Difference 5 – 2 = 3 5 > 2 5 > 3

Minuend = Difference + Subtrahend

? - 2 = 3 ? = 3 + 2 ? = 5

Subtrahend = Minuend – Difference

5 - ? = 3 ? = 5 – 3 ? = 2

Review 1 B

Complete the tables

5

	a	b	c	d	e
Minuend		81100	81100		85540
Subtrahend	19190		17370	16460	
Difference	61910	62820		64640	65950

6

Minuend		82210	82210		85540
Subtrahend	19290		17470	16560	
Difference	62920	63830		65650	66860

7

	a	b	c	d	e
Minuend	?	83320	83320		85540
Subtrahend	19390		17570	16660	
Difference	63930	64840		66660	68680

a) ? - 19390 = 63930

8

Minuend	?	84430	84430	?	85540
Subtrahend	19490	?	16760	17670	?
Difference	64940	65850	?	66760	67770

Grade 6 Chapter 5 Lesson 4
Addition and subtraction 3
Rules for addition

Addend + Addend = Sum   2 + 3 = 5   5 > 2   5 > 3
First Addend = Sum – Second Addend
? + 3 = 5   ? = 5 – 3   ? = 2
Second Addend = Sum – First Addend
2 + ? = 5   ? = 5 – 2   ? = 3

Review 2 A
Complete the tables

9

	a	b	c	d	e
Addend		69790	68980		63930
Addend	26960		29890	28580	
Sum	96650	97760		94430	93320

10

Addend		68880	69890		64840
Addend	29690		28980	29490	
Sum	96650	97760		94430	93320

11

	a	b	c	d	e
Addend		67970	69990		65650
Addend	27870	?	29990	27570	
Sum	96650	97760		93320	92210

b) 67970 + ? = 97760

12

Addend	?	65550	65950	?	64740
Addend	28780	?	29590	28480	?
Sum	96650	91100	?	93320	92210

Grade 6 Chapter 5 Lesson 4
Addition and subtraction 3
Rules for subtraction

Minuend – Subtrahend = Difference   5 – 2 = 3   5 > 2   5 > 3
Minuend = Difference + Subtrahend
? - 2 = 3   ? = 3 + 2   ? = 5
Subtrahend = Minuend – Difference
5 - ? = 3   ? = 5 – 3   ? = 2

Review 2 B

Complete the tables

 # 13

	a	b	c	d	e
Minuend		97760	98870		93320
Subtrahend	29690		28980	25850	
Difference	66960	67970		68580	69390

 # 14

Minuend		97760	98870		93320
Subtrahend	26960		29890	24940	
Difference	69690	68880		69490	68480

 # 15

	a	b	c	d	e
Minuend		97760	99980		92210
Subtrahend	27870	?	29990	25750	
Difference	68780	69790		67570	66560

b) 97760 - ? = 69790

 # 16

Minuend	?	91100	95540	?	92210
Subtrahend	27870	?	25950	24840	?
Difference	68780	65550	?	68480	67470

Grade 6 Chapter 5 Lesson 5
Multiplication 1
Rules for multiplication

Factor × Factor = Product $2 \times 3 = 6$ $6 > 2$ $6 > 3$

First Factor = Product ÷ Second Factor

$? \times 3 = 6$ $? = 6 \div 3$ $? = 2$

Second Factor = Product ÷ First Factor

$2 \times ? = 6$ $? = 6 \div 2$ $? = 3$

Review 1 A
Find unknown number

 # 1

a) ? × 32 = 12800

b) 32 × ? = 12832

c) ? × 32 = 12864

d) 32 × ? = 12896

e) ? × 32 = 12928

 # 2

a) ? × 33 = 13365

b) 33 × ? = 13398

c) ? × 33 = 13431

d) 33 × ? = 13464

e) ? × 33 = 13497

 # 3

a) ? × 34 = 13906

b) 34 × ? = 13872

c) ? × 34 = 13838

d) 34 × ? = 13804

e) ? × 34 = 13770

 # 4

a) ? × 35 = 14140

b) 35 × ? = 14105

c) ? × 35 = 14070

d) 35 × ? = 14035

c) ? × 35 = 14000

Grade 6 Chapter 5 Lesson 5
Multiplication 1
Rules for multiplication

Factor × Factor = Product $2 \times 3 = 6$ $6 > 2$ $6 > 3$

First Factor = Product ÷ Second Factor
$? \times 3 = 6$ $? = 6 \div 3$ $? = 2$

Second Factor = Product ÷ First Factor
$2 \times ? = 6$ $? = 6 \div 2$ $? = 3$

Review 1 B
Find unknown number

 # 5
a) $? \times 36 = 14580$
b) $36 \times ? = 14616$
c) $? \times 36 = 14652$
d) $36 \times ? = 14688$
e) $? \times 36 = 14724$

 # 6
a) $? \times 37 = 14948$
b) $37 \times ? = 14911$
c) $? \times 37 = 14874$
d) $37 \times ? = 14837$
e) $? \times 37 = 14800$

 # 7
a) $? \times 38 = 15200$
b) $38 \times ? = 15238$
c) $? \times 38 = 15276$
d) $38 \times ? = 15314$
c) $? \times 38 = 15352$

 # 8
a) $? \times 39 = 15795$
b) $39 \times ? = 15834$
c) $? \times 39 = 15873$
d) $39 \times ? = 15912$
c) $? \times 39 = 15951$

Grade 6 Chapter 5 Lesson 6
Multiplication 2
Rules for multiplication

Factor × Factor = Product $2 \times 3 = 6$ $6 > 2$ $6 > 3$

First Factor = Product ÷ Second Factor
$? \times 3 = 6$ $? = 6 \div 3$ $? = 2$

Second Factor = Product ÷ First Factor
$2 \times ? = 6$ $? = 6 \div 2$ $? = 3$

Review 1 A
Find unknown number

1

	a	b	c	d	e
Factor		42		42	
Factor	42		42		42
Product	21000	21042	21084	21126	21168

2

Factor		43		43	
Factor	43		43		43
Product	21715	21758	21801	21844	21887

3

e) $? \times 44 = 22220$

					e
Factor		44		44	?
Factor	44		44		44
Product	22396	22352	22308	22264	22220

4

Factor	?	45	?	45	?
Factor	45	?	45	?	45
Product	22680	22635	22590	22545	22500

Grade 6 Chapter 5 Lesson 6
Multiplication 2
Rules for multiplication

Factor × Factor = Product $2 \times 3 = 6$ $6 > 2$ $6 > 3$

First Factor = Product ÷ Second Factor
$? \times 3 = 6$ $? = 6 \div 3$ $? = 2$

Second Factor = Product ÷ First Factor
$2 \times ? = 6$ $? = 6 \div 2$ $? = 3$

Review 1 B
Find unknown number

 # 5

	a	b	c	d	e
Factor		46		46	
Factor	46		46		46
Product	23230	23276	23322	23368	23414

 # 6

Factor		47		47	
Factor	47		47		47
Product	23688	23641	23594	23547	23500

 # 7

	a	b	c	d	e
Factor		48		48	
Factor	48	?	48		48
Product	24000	24048	24096	24144	24192

b) 48 × ? = 24048

 # 8

Factor	?	49	?	49	?
Factor	49	?	49	?	49
Product	24745	24794	24843	24892	24941

Grade 6 Chapter 5 Lesson 7
Division 1
Rules for division

Dividend ÷ Divisor = Quotient $6 \div 2 = 3$ $6 > 2$ $6 > 3$

Divisor = Dividend ÷ Quotient

$6 \div ? = 3$ $? = 6 \div 3$ $? = 2$

Dividend = Quotient × Divisor

$? \div 2 = 3$ $? = 3 \times 2$ $? = 6$

Review 1 A
Find unknown number

 # 1

a) $43400 \div ? = 62$

b) $? \div 62 = 701$

c) $43524 \div ? = 62$

d) $? \div 62 = 703$

e) $43648 \div 62 = ?$

 # 2

a) $44415 \div ? = 63$

b) $? \div 63 = 706$

c) $44541 \div ? = 63$

d) $? \div 63 = 708$

e) $44667 \div 63 = ?$

 # 3

a) $45376 \div ? = 64$

b) $? \div 64 = 708$

c) $45248 \div ? = 64$

d) $? \div 64 = 706$

e) $45120 \div 64 = ?$

 # 4

a) $45760 \div ? = 65$

b) $? \div 65 = 703$

c) $45630 \div ? = 65$

d) $? \div 65 = 701$

e) $45500 \div 65 = ?$

Grade 6 Chapter 5 Lesson 7
Division 1
Rules for division

Dividend ÷ Divisor = Quotient $6 \div 2 = 3$ $6 > 2$ $6 > 3$

Divisor = Dividend ÷ Quotient
$6 \div ? = 3$ $? = 6 \div 3$ $? = 2$

Dividend = Quotient × Divisor
$? \div 2 = 3$ $? = 3 \times 2$ $? = 6$

Review 1 B
Find unknown number

5
a) $46530 \div ? = 66$
b) $? \div 66 = 706$
c) $46662 \div ? = 66$
d) $? \div 66 = 708$
e) $46794 \div 66 = ?$

6
a) $47168 \div ? = 67$
b) $? \div 67 = 703$
c) $47034 \div ? = 67$
d) $? \div 67 = 701$
e) $46900 \div 67 = ?$

7
a) $47600 \div ? = 68$
b) $? \div 68 = 701$
c) $47736 \div ? = 68$
d) $? \div 68 = 703$
e) $47872 \div 68 = ?$

8
a) $48645 \div ? = 69$
b) $? \div 69 = 706$
c) $48783 \div ? = 69$
d) $? \div 69 = 708$
e) $48921 \div 69 = ?$

Grade 6 Chapter 5 Lesson 8
Division 2
Rules for division

Dividend ÷ Divisor = Quotient $6 \div 2 = 3$ $6 > 2$ $6 > 3$

Divisor = Dividend ÷ Quotient

$6 \div ? = 3$ $? = 6 \div 3$ $? = 2$

Dividend = Quotient × Divisor

$? \div 2 = 3$ $? = 3 \times 2$ $? = 6$

Review 1 A
Find unknown number

1

	a	b	c	d	e
Dividend		57672	57744		57888
Divisor	72		72	803	72
Quotient	800	72		72	

2

Dividend		58838	58911		59057
Divisor	73		73	808	73
Quotient	805	73		73	

3

		b			
Dividend		59792	59718		59570
Divisor	74	?	74	806	74
Quotient	809	74		74	

b) $59792 \div ? = 74$

4

Dividend	?	60225	60150	?	60000
Divisor	75	?	75	801	75
Quotient	804	75	?	75	?

Grade 6 Chapter 5 Lesson 8
Division 2
Rules for division

Dividend ÷ Divisor = Quotient 6 ÷ 2 = 3 6 > 2 6 > 3

Divisor = Dividend ÷ Quotient
6 ÷ ? = 3 ? = 6 ÷ 3 ? = 2

Dividend = Quotient × Divisor
? ÷ 2 = 3 ? = 3 × 2 ? = 6

Review 1 B
Find unknown number

5

	a	b	c	d	e
Dividend		61256	61332		61484
Divisor	76		76	808	76
Quotient	805	76		76	

6

Dividend		61831	61754		61600
Divisor	77		77	801	77
Quotient	804	77		77	

7 b) 62478 ÷ ? = 78

	a	b	c	d	e
Dividend		62478	62556		62712
Divisor	78	?	78	803	78
Quotient	800	78		78	

8

Dividend	?	63674	63753	?	63911
Divisor	79	?	79	808	79
Quotient	805	79	?	79	?

Grade 6 Chapter 5 Lesson 9
Division 3
Rules for division

Dividend ÷ Divisor = Quotient $6 \div 2 = 3$ $6 > 2$ $6 > 3$

Divisor = Dividend ÷ Quotient

$6 \div ? = 3$ $? = 6 \div 3$ $? = 2$

Dividend = Quotient × Divisor

$? \div 2 = 3$ $? = 3 \times 2$ $? = 6$

Review 2 A
Find unknown number

\# 1

	a	b	c	d	e
Dividend		73882	73964		74128
Divisor	82		82	903	82
Quotient	900	82		82	

\# 2

Dividend		75198	75281		75447
Divisor	83		83	908	83
Quotient	905	83		83	

\# 3 a

Dividend	?	76272	76188		76020
Divisor	84		84	906	84
Quotient	909	84		84	

a) $? \div 84 = 909$

\# 4

Dividend	?	76755	76670	?	76500
Divisor	85	?	85	901	85
Quotient	904	85	?	85	?

Grade 6 Chapter 5 Lesson 9
Division 3
Rules for division

Dividend ÷ Divisor = Quotient 6 ÷ 2 = 3 6 > 2 6 > 3

Divisor = Dividend ÷ Quotient
6 ÷ ? = 3 ? = 6 ÷ 3 ? = 2

Dividend = Quotient × Divisor
? ÷ 2 = 3 ? = 3 × 2 ? = 6

Review 2 B
Find unknown number

#5

	a	b	c	d	e
Dividend		77916	78002		78174
Divisor	86		86	908	86
Quotient	905	86		86	

#6

Dividend		78561	78474		78300
Divisor	87		87	901	87
Quotient	904	87		87	

#7

	a	b	c	d	e
Dividend		79288	79376	?	79552
Divisor	88		88	903	88
Quotient	900	88		88	

d) ? ÷ 903 = 88

#8

Dividend	?	80634	80723	?	80901
Divisor	89	?	89	908	89
Quotient	905	89	?	89	?

Grade 6 Chapter 5 Lesson 10
Multiplication and Division 1
Review 1 A

	# 1	# 2	# 3	# 4	# 5
Factor		6		8	5006
Factor	5		7		9
Product	5010	12018	21028	32040	

	# 6	# 7	# 8	# 9	# 10
Factor		87		65	506
Factor	98	?	76		54
Product	9996	17661	23104	26325	

7) 87 × ? = 17661

	# 11	# 12	# 13	# 14	# 15
Dividend	30035	42048	56063	72064	
Divisor	5	?	7		9
Quotient		6		8	8007

12) 42048 ÷ ? = 6

	# 16	# 17	# 18	# 19	# 20
Dividend	58272	60180	59866	57204	?
Divisor	96	?	74	?	52
Quotient	?	85	?	63	807

Review 1 B

	# 1	# 2	# 3	# 4	# 5
Factor		8		6	5007
Factor	9		7		5
Product	9027	16032	21035	24036	

	# 6	# 7	# 8	# 9	# 10
Factor		86		64	507
Factor	97	?	75		53
Product	9991	17544	22875	25984	

7) 86 × ? = 17544

	# 11	# 12	# 13	# 14	# 15
Dividend	54072	56072	56042	54042	
Divisor	9	?	7		5
Quotient		8		6	8006

12) 54072 ÷ ? = 8

	# 16	# 17	# 18	# 19	# 20
Dividend	57760	59556	58838	56234	?
Divisor	95	?	73	?	51
Quotient	?	84	?	62	804

Grade 6 Chapter 5 Lesson 10
Multiplication and Division 1
Review 2 A

	# 1	# 2	# 3	# 4	# 5
Factor	9		7		5006
Factor		8		6	5
Product	9018	16024	21028	24030	

	# 6	# 7	# 8	# 9	# 10
Factor		80		60	506
Factor	90	?	70		50
Product	9180	16240	21280	24300	

7) 80 × ? = 16240

	# 11	# 12	# 13	# 14	# 15
Dividend	24028	21024	16018	9008	
Divisor	4	?	2		1
Quotient		3		1	8007

12) 21024 ÷ ? = 3

	# 16	# 17	# 18	# 19	# 20
Dividend	24280	21240	16180	9080	?
Divisor	40	?	20	?	10
Quotient	?	30	?	10	807

Review 2 B

	# 1	# 2	# 3	# 4	# 5
Factor	9		7		5007
Factor		8		6	6
Product	9027	16032	21035	24036	

	# 6	# 7	# 8	# 9	# 10
Factor		80		60	507
Factor	90	?	70		50
Product	9270	16320	21350	24360	

7) 80 × ? = 16320

	# 11	# 12	# 13	# 14	# 15
Dividend	24032	21027	16014	9007	
Divisor	4	?	2		1
Quotient		3		1	8006

12) 21027 ÷ ? = 3

	# 16	# 17	# 18	# 19	# 20
Dividend	24320	21270	16140	9090	?
Divisor	40	?	20	?	10
Quotient	?	30	?	10	809

Grade 6 Chapter 5 Lesson 11
Multiplication and Division 2
Review 1 A

	# 1	# 2	# 3	# 4	# 5
Factor		8		6	5008
Factor	9		7		5
Product	9036	16040	21042	24042	

	# 6	# 7	# 8	# 9	# 10
Factor		80		60	508
Factor	90	?	70		50
Product	9360	16400	21420	24420	

7) 80 × ? = 16400

	# 11	# 12	# 13	# 14	# 15
Dividend	24036	21018	16012	9006	
Divisor	4	?	2		1
Quotient		3		1	8005

12) 21018 ÷ ? = 3

	# 16	# 17	# 18	# 19	# 20
Dividend	24360	21150	16120	9070	?
Divisor	40	?	20	?	10
Quotient	?	30	?	10	805

Review 1 B

	# 1	# 2	# 3	# 4	# 5
Factor		8		6	5009
Factor	9		7		5
Product	9045	16048	21049	24048	

	# 6	# 7	# 8	# 9	# 10
Factor		80		60	509
Factor	90	?	70		50
Product	9450	16480	21490	24480	

7) 80 × ? = 16480

	# 11	# 12	# 13	# 14	# 15
Dividend	24032	21015	16010	9005	
Divisor	4	?	2		1
Quotient		3		1	8009

12) 21015 ÷ ? = 3

	# 16	# 17	# 18	# 19	# 20
Dividend	24200	21180	16100	9060	?
Divisor	40	?	20	?	10
Quotient	?	30	?	10	806

Grade 6 Chapter 5 Lesson 11
Multiplication and Division 2
Review 2 A

	# 1	# 2	# 3	# 4	# 5
Factor	9		7		5008
Factor		8		6	6
Product	9045	16048	21049	24048	

	# 6	# 7	# 8	# 9	# 10
Factor		80		60	507
Factor	90	?	70		50
Product	9450	16480	21490	24480	

7) 80 × ? = 16480

	# 11	# 12	# 13	# 14	# 15
Dividend	24032	21021	16012	9005	
Divisor	4	?	2		1
Quotient		3		1	8004

12) 21021 ÷ ? = 3

	# 16	# 17	# 18	# 19	# 20
Dividend	24320	21210	16120	9050	?
Divisor	40	?	20	?	10
Quotient	?	30	?	10	804

Review 2 B

	# 1	# 2	# 3	# 4	# 5
Factor		8		6	5009
Factor	9		7		6
Product	9054	16056	21056	24054	

	# 6	# 7	# 8	# 9	# 10
Factor		80		60	506
Factor	90	?	70		50
Product	9540	16560	21560	24540	

7) 80 × ? = 16560

	# 11	# 12	# 13	# 14	# 15
Dividend	24028	21018	16010	9004	
Divisor	4	?	2		1
Quotient		3		1	8003

12) 21018 ÷ ? = 3

	# 16	# 17	# 18	# 19	# 20
Dividend	24280	21210	16080	9050	?
Divisor	40	?	20	?	10
Quotient	?	30	?	10	806

Grade 6 Chapter 5 Lesson 12
Order of operations with numbers 1
Review 1 A

1 Subtract

$$\begin{array}{r} 2000 \\ \underline{-\ 1092} \end{array}$$

2 Multiply

$$\begin{array}{r} 456 \\ \underline{\times\ 7} \end{array}$$

3

$$\begin{array}{r} 72 \\ \underline{\times\ 65} \end{array}$$

4 Divide $4560 \div 8$

5 $8091 \div 87$

Evaluate

6 $27 \times 54 \div 6$

7 $1728 \div 8 \times 9$

8 $3456 \div 9 \div 8$

9 $1000 - 85 \times 9$

10 $6192 - 6102 \div 9$

Find the unknown number

	# 11	# 12	# 13	# 14	# 15
Factor	?	46	47	?	49
Factor	39	?	37	36	?
Product	1755	1748	?	1728	1715

11) $? \times 39 = 1755$

	# 16	# 17	# 18	# 19	# 20
Dividend	?	5168	5159	?	5135
Divisor	69	?	67	66	?
Quotient	75	76	?	78	79

16) $? \div 69 = 75$

Grade 6 Chapter 5 Lesson 12
Order of operations with numbers 1

Review 1 B

1 Subtract 3000 − 2191

2 Multiply 567 × 8

3 82 × 75

4 Divide 5670 ÷ 9

5 7332 ÷ 78

Evaluate

6 27 × 72 ÷ 9
7 1458 ÷ 9 × 8
8 3888 ÷ 8 ÷ 9
9 1000 – 95 × 7
10 4616 - 4536 ÷ 8

Find the unknown number

	# 11	# 12	# 13	# 14	# 15
Factor	?	36	37	?	39
Factor	49	?	47	46	?
Product	1715	1728	?	1748	1755

15) 39 × ? = 1755

	# 16	# 17	# 18	# 19	# 20
Dividend	?	3828	3819	?	3795
Divisor	59	?	57	56	?
Quotient	65	66	?	68	69

20) 3795 ÷ ? = 69

Grade 6 Chapter 5 Lesson 13
Order of operations with numbers 2

Review 5 A

1 Subtract

$$\begin{array}{r} 4000 \\ \underline{-\ 2182} \end{array}$$

2 Multiply

$$\begin{array}{r} 678 \\ \underline{\times\ 7} \end{array}$$

3

$$\begin{array}{r} 83 \\ \underline{\times\ 75} \end{array}$$

4 Divide $6030 \div 9$

5 $7332 \div 78$

Evaluate

6 $34 \times 63 \div 7$
7 $1664 \div 8 \times 9$
8 $3312 \div 9 \div 8$
9 $1000 - 75 \times 9$
10 $9306 - 7101 \div 9$

Find the unknowns numbers

	# 11	# 12	# 13	# 14	# 15
Factor	?	46	47	?	49
Factor	49	?	47	46	?
Product	2205	2208	?	2208	2205

	# 16	# 17	# 18	# 19	# 20
Dividend	?	4408	4389	?	4345
Divisor	59	?	57	56	?
Quotient	75	76	?	78	79

Grade 6 Chapter 5 Lesson 13
Order of operations with numbers 2

Review 5 B

1 Subtract $5000 - 3292$

2 Multiply 789×8

3 73×65

4 Divide $5120 \div 8$

5 $6794 \div 79$

Evaluate

6 $37 \times 72 \div 9$
7 $1648 \div 8 \times 9$
8 $4608 \div 9 \div 8$
9 $1000 - 85 \times 9$
10 $9738 - 6111 \div 9$

Find unknown number

	# 11	# 12	# 13	# 14	# 15
Factor	?	36	37	?	39
Factor	59	?	57	56	?
Product	2065	2088	?	2128	2145

	# 16	# 17	# 18	# 19	# 20
Dividend	?	3648	3619	?	3555
Divisor	49	?	47	46	?
Quotient	75	76	?	78	79

Grade 6 Chapter 5 Lesson 14
Test 5 A
Find unknown number

1 $1 \times ? = 5006$
2 $? \times 1 = 6007$
3 $2 \times ? = 14016$
4 $? \times 3 = 24027$
5 $4 \times 9009 = ?$

	# 6	# 7	# 8	# 9	# 10
Factor		65		85	95
Factor	55		75		405
Product	5500	6630	15225	25840	

11 $? \div 9 = 9008$
12 $64056 \div ? = 8$
13 $? \div 7 = 7006$
14 $36030 \div ? = 6$
15 $25020 \div 5 = ?$

	# 16	# 17	# 18	# 19	# 20
Dividend		9045		28035	40500
Divisor	10		25		45
Quotient	5040	15	702	35	

Grade 6 Chapter 5 Lesson 14
Test 5 B
Find unknown number

1 $5 \times ? = 5000$
2 $? \times 6 = 6012$
3 $7 \times ? = 14021$
4 $? \times 8 = 24032$
5 $9 \times 4005 = ?$

	# 6	# 7	# 8	# 9	# 10
Factor		15		35	909
Factor	10		25		45
Product	5060	9105	17700	28315	

11 $16012 \div ? = 4$
12 $? \div 3002 = 3$
13 $4002 \div ? = 2$
14 $? \div 2003 = 1$
15 $1002 \div 1 = ?$

	# 16	# 17	# 18	# 19	# 20
Dividend		6630		25840	38475
Divisor	55		75		95
Quotient	100	65	203	85	

Grade 6 Chapter 5 Lesson 15
Final Exam 6 A
(Basic Problems) # 1 - # 10
Many digit numbers
Show all steps for full credits not just answers
Each problem is worth 5 points
Evaluate

1) $\begin{array}{r} 700000 \\ \underline{-\ 234567} \end{array}$

2) $\begin{array}{r} 765 \\ \underline{\times 604} \end{array}$

3) $\begin{array}{r} 706 \\ \underline{\times 605} \end{array}$

4) $193884 \div 321$

5) $243714 \div 302$

6) $240240 \div 80 \times 3$

7) $7490 - 830 \times 3$

8) $56049 - 49042 \div 7$

9) $(30 \times 560) \div 8$

10) $289 \div (93 - 76)$

Grade 6 Chapter 5 Lesson 16
Final Exam 6 B (Basic Problems) # 1 - # 10
Many digit numbers

Show all steps for full credits not just answers. Each problem is worth 5 points

Evaluate

1) $\begin{array}{r} 600000 \\ \underline{-\ 123456} \end{array}$

2) $\begin{array}{r} 654 \\ \underline{\times 503} \end{array}$

3) $\begin{array}{r} 605 \\ \underline{\times 403} \end{array}$

4) $105630 \div 210$

5) $182508 \div 201$

6) $180180 \div 90 \times 2$

7) $6840 - 920 \times 2$

8) $42036 - 36030 \div 6$

9) $(20 \times 720) \div 9$

10) $256 \div (81 - 65)$

Grading Scale: A =10 points, B=9points, C=8 points, D=7 points, F=0-6 points

Lesson 16 Final Exam 6 B (Basic Problems) # 1 - # 10

1)	476544	2)	328962	3)	243815	4)	503	5)	908
6)	4004	7)	5000	8)	36031	9)	1600	10)	16

Grade 6 Answers Chapter 5

Lesson 2 Review 1 A

	a	b		a	b
#1	1919	1393	#11	3110	3939
#2	2828	1484	#12	4110	4848
#3	3737	1575	#13	5110	5757
#4	4646	1696	#14	6110	6969
#5	5959	1797	#15	7554	7979
#6	2929	1494	#16	4221	4949
#7	3838	1585	#17	5221	5858
#8	4747	1676	#18	6221	6767
#9	5656	1787	#19	7221	7878
#10	6868	898	#20	8554	8989

Lesson 2 Review 1 B

	a	b		c	d
# 1	49690	29690	# 11	76650	28780
# 2	48780	28780	# 12	76650	27870
# 3	49790	29790	# 13	77760	27970
# 4	48880	25550	# 14	77760	25550
# 5	48980	29990	# 15	78870	29990
# 6	49890	29590	# 16	78870	25950
# 7	45850	27570	# 17	74430	25750
# 8	44940	28480	# 18	74430	24840
# 9	43930	26560	# 19	73320	25650
# 10	44840	27470	# 20	73320	24740

Lesson 3

Review 1 A	a	b	c	d	e
# 1	69190	12820	81100	66460	15950
# 2	69290	13830	82210	66560	16860
# 3	69390	14840	83320	66660	18680
# 4	69490	15850	84430	67670	17770
Review 1 B					
# 5	81100	18280	63730	81100	19590
# 6	82210	18380	64740	82210	18680
# 7	83320	18480	65750	83320	16860
# 8	84430	18580	67670	84430	17770

Grade 6 Answers Chapter 5

Lesson 4 Review 2 A

	a	b	c	d	e
# 9	69690	27970	98870	65850	29390
# 10	66960	28880	98870	64940	28480
# 11	68780	29790	99980	65750	26560
# 12	67870	25550	95540	64840	27470

Review 2 B

	a	b	c	d	e
# 13	96650	29790	69890	94430	23930
# 14	96650	28880	68980	94430	24840
# 15	96650	27970	69990	93320	25650
# 16	96650	25550	69590	93320	24740

Lesson 5

Review 1 A

	# 1	# 2	# 3	# 4
a)	400	405	409	404
b)	401	406	408	403
c)	402	407	407	402
d)	403	408	406	401
e)	404	409	405	400

Review 1 B

	# 5	# 6	# 7	# 8
a)	405	404	400	405
b)	406	403	401	406
c)	407	402	402	407
d)	408	401	403	408
e)	409	400	404	409

Lesson 6

Review 1 A

	a	b	c	d	e
# 1	500	501	502	503	504
# 2	505	506	507	508	509
# 3	509	508	507	506	505
# 4	504	503	502	501	500

Review 1 B

	a	b	c	d	e
# 5	505	506	507	508	509
# 6	504	503	502	501	500
# 7	500	501	502	503	504
# 8	505	506	507	508	509

Lesson 7

Review 1 A

	# 1	# 2	# 3	# 4
a)	700	705	709	704
b)	43462	44478	45312	45695
c)	702	707	707	702
d)	43586	44604	45184	45565
e)	704	709	705	700

Grade 6 Answers Chapter 5

Lesson 7 Review 1 B

#5		#6		#7		#8	
a)	705	a)	704	a)	700	a)	705
b)	46596	b)	47101	b)	47668	b)	48714
c)	707	c)	702	c)	702	c)	707
d)	46728	d)	46967	d)	47804	d)	48852
e)	709	e)	700	e)	704	e)	709

Lesson 8

Review 1 A

	a	b	c	d	e
# 1	57600	801	802	57816	804
# 2	58765	806	807	58984	809
# 3	59866	808	807	59644	805
# 4	60300	803	802	60075	800

Review 1 B

	a	b	c	d	e
# 5	61180	806	807	61408	809
# 6	61908	803	802	61677	800
# 7	62400	801	802	62634	804
# 8	63595	806	807	63832	809

Lesson 9

		a	b	c	d	e
Review 1 A	# 1	73800	901	902	74046	904
	# 2	75115	906	907	75364	909
	# 3	76356	908	907	76104	905
	# 4	76840	903	902	76585	900
Review 1 B	# 5	77830	906	907	78088	909
	# 6	78648	903	902	78387	900
	# 7	79200	901	902	79464	904
	# 8	80545	906	907	80812	909

Grade 6 Answers Chapter 5

Lesson 10

Review 1 A

# 1	1002	# 6	102	# 11	6007	# 16	607
# 2	2003	# 7	203	# 12	7008	# 17	708
# 3	3004	# 8	304	# 13	8009	# 18	809
# 4	4005	# 9	405	# 14	9008	# 19	908
# 5	45054	# 10	27324	# 15	72063	# 20	41964

Review 1 B

# 1	1003	# 6	103	# 11	6008	# 16	608
# 2	2004	# 7	204	# 12	7009	# 17	709
# 3	3005	# 8	305	# 13	8006	# 18	806
# 4	4006	# 9	406	# 14	9007	# 19	907
# 5	25035	# 10	26871	# 15	40030	# 20	41004

Review 2 A

# 1	1002	# 6	102	# 11	6007	# 16	607
# 2	2003	# 7	203	# 12	7008	# 17	708
# 3	3004	# 8	304	# 13	8009	# 18	809
# 4	4005	# 9	405	# 14	9008	# 19	908
# 5	25030	# 10	25300	# 15	8007	# 20	8070

Review 2 B

# 1	1003	# 6	103	# 11	6008	# 16	608
# 2	2004	# 7	204	# 12	7009	# 17	709
# 3	3005	# 8	305	# 13	8007	# 18	807
# 4	4006	# 9	406	# 14	9007	# 19	909
# 5	30042	# 10	25350	# 15	8006	# 20	8090

Grade 6 Answers Chapter 5

Lesson 11

Review 1 A

# 1	1004	# 6	104	# 11	6009	# 16	609
# 2	2005	# 7	205	# 12	7006	# 17	705
# 3	3006	# 8	306	# 13	8006	# 18	806
# 4	4007	# 9	407	# 14	9006	# 19	907
# 5	25040	#10	25400	# 15	8005	# 20	8050

Review 1 B

# 1	1005	# 6	105	# 11	6008	# 16	605
# 2	2006	# 7	206	# 12	7005	# 17	706
# 3	3007	# 8	307	# 13	8005	# 18	805
# 4	4008	# 9	408	# 14	9005	# 19	906
# 5	25045	# 10	25450	# 15	8009	# 20	8060

Review 2 A

# 1	1005	# 6	105	# 11	6008	# 16	608
# 2	2006	# 7	206	# 12	7007	# 17	707
# 3	3007	# 8	307	# 13	8006	# 18	806
# 4	4008	# 9	408	# 14	9005	# 19	905
# 5	30048	#10	25350	# 15	8004	# 20	8040

Review 2 B

# 1	1006	# 6	106	# 11	6007	# 16	607
# 2	2007	# 7	207	# 12	7006	# 17	707
# 3	3008	# 8	308	# 13	8005	# 18	804
# 4	4009	# 9	409	# 14	9004	# 19	905
# 5	30054	# 10	25300	# 15	8003	# 20	8060

Lesson 12

Review 1 A

#1	908	#2	3192	#3	4680	#4	570	#5	93
#6	243	#7	1944	#8	48	#9	235	#10	5514
#11	45	#12	38	#13	1739	#14	48	#15	35
#16	5175	#17	68	#18	77	#19	5148	#20	65

Review 1 B

#1	809	#2	4536	#3	6150	#4	630	#5	94
#6	216	#7	1296	#8	54	#9	335	#10	4049
#11	35	#12	48	#13	1739	#14	38	#15	45
#16	3835	#17	58	#18	67	#19	3808	#20	55

Grade 6 Answers Chapter 5

Lesson 13

Review 5 A

#1	1818	#2	4746	#3	6225	#4	670	#5	94
#6	306	#7	1872	#8	46	#9	325	#10	8517
#11	45	#12	48	#13	2209	#14	48	#15	45
#16	4425	#17	58	#18	77	#19	4368	#20	55

Review 5 B

#1	1708	#2	6312	#3	4745	#4	640	#5	86
#6	296	#7	1854	#8	64	#9	235	#10	9059
#11	35	#12	58	#13	2109	#14	38	#15	55
#16	3675	#17	48	#18	77	#19	3588	#20	45

Lesson 14 Test 5 A

# 1	5006	# 6	100	# 11	81072	# 16	504
# 2	6007	# 7	102	# 12	8007	# 17	603
# 3	7008	# 8	203	# 13	49042	# 18	17550
# 4	8009	# 9	304	# 14	6005	# 19	801
# 5	36036	# 10	38475	# 15	5004	# 20	900

Test 5 B

# 1	1000	# 6	506	# 11	4003	# 16	5500
# 2	1002	# 7	607	# 12	9006	# 17	102
# 3	2003	# 8	708	# 13	2001	# 18	15525
# 4	3004	# 9	809	# 14	2003	# 19	304
# 5	36045	# 10	40905	# 15	1002	# 20	405

Lesson 15 Final Exam 6 A (Basic Problems) # 1 - # 10

1)	465433	2)	462060	3)	427130	4)	604	5)	807
6)	9009	7)	5000	8)	49043	9)	2100	10)	17

Lesson 16 Final Exam 6 B (Basic Problems) # 1 - # 10

1)	476544	2)	328962	3)	243815	4)	503	5)	908
6)	4004	7)	5000	8)	36031	9)	1600	10)	16

Grade 6 Chapter 5 Summary 1
Order of Operations with Numbers 1

Addition and Subtraction Numbers
Addition of numbers
To add numbers, add in any order

Example 1 7000 + 35000 + 3000 = 7000 + 38000 = 45000

7000 + 35000 + 3000 = 42000 + 3000 = 45000

7000 + 35000 + 3000=7000 + 3000 + 35000=10000 + 35000=45000

$$\begin{array}{r} 35000 \\ +\ 7000 \\ \underline{3000} \\ 45000 \end{array}$$

Addition and Subtraction of numbers
To add and subtract numbers work from left to right

Example 2 7000 + 35000 - 3000 = 42000 - 3000 = 39000

1) $\begin{array}{r} 35000 \\ \underline{+\ 7000} \\ 42000 \end{array}$ 2) $\begin{array}{r} 42000 \\ \underline{-3000} \\ 39000 \end{array}$

Subtraction and Addition of numbers
To subtract and add numbers work from left to right

Example 3 35000 – 3000 + 7000 = 32000 + 7000 = 39000

1) $\begin{array}{r} 35000 \\ \underline{-\ 3000} \\ 32000 \end{array}$ 2) $\begin{array}{r} 32000 \\ \underline{+7000} \\ 39000 \end{array}$

Subtraction of numbers
To subtract numbers, work from left to right

Example 4 35000 – 3000 - 7000 = 32000 - 7000 = 25000

1) $\begin{array}{r} 35000 \\ \underline{-\ 3000} \\ 32000 \end{array}$ 2) $\begin{array}{r} 32000 \\ \underline{-\ 7000} \\ 25000 \end{array}$

Grade 6 Chapter 5 Summary 1
Order of Operations with Numbers 2

2) Multiplication numbers: for memory <u>(x, x, any)</u>
To multiply numbers, work in any order
$9\times 8 \times 7 = 72 \times 7 = 504$ $\quad$ $9 \times 8 \times 7 = 9 \times 56 = 504$

2) Work from left to right: for memory <u>(L → R)</u>

d) Multiplication and Division numbers: (× or ÷)
To multiply and divide numbers work from left to right (L → R)
$9\times 6 \div 3 = 54 \div 3 = 18$ $\quad$ $4 \times 81 \div 9 = 324 \div 9 = 36$

e) Division and Multiplication numbers: (÷ or ×)
To divide and multiply numbers work from left to right <u>(L → R)</u>
$25 \div 6 \times 2 = 4 \times 2 = 8$ $\quad$ $81 \div 9 \times 4 = 9 \times 4 = 36$

f) Division numbers: (÷, ÷)
To divide numbers, work from left to right <u>(L → R)</u>
$25 \div 4 \div 2 = 6 \div 2 = 3$ $\quad$ $810\div 9\div 3 = 90\div 9 = 10$

3) Do Multiplication or Division before Addition or Subtraction
for memory (× or ÷) before (+ or -)
e) Multiplication and Addition numbers: 1) × 2) +
To multiply and add numbers always first multiply than add
$9\times 6 + 9 = 54 + 9 = 63$ $\quad$ $9 + 9 \times 9 = 9 + 81 = 90$

f) Multiplication and Subtraction numbers: 1) × 2) −
To multiply and subtract numbers always first multiply than subtract
$9\times 9 - 9 = 81 - 9 = 72$ $\quad$ $90 - 9 \times 9 = 90 - 81 = 9$

g) Division and Addition numbers: 1) ÷ 2) +
To divide and add numbers always first divide than add
$51 \div 5 + 39 = 10 + 39 = 49$ $\quad$ $88 + 171 \div 9 = 88 + 19 = 107$

h) Division and Subtraction numbers: 1) ÷ 2) −
To divide and subtract numbers always first divide than subtract
$108 \div 6 - 9 = 18 - 9 = 9$ $\quad$ $500 - 126 \div 7 = 500 - 18 = 482$

Grade 6 Chapter 5 Summary 1
Order of Operations with Numbers 3

Multiplication of Numbers

To multiply numbers, work in any order

$405 \times 8 \times 9 = 3240 \times 9 = 29160$ $\quad$ $405 \times 8 \times 9 = 405 \times 72 = 29160$

$506 \times 9 \times 8 = 4554 \times 8 = 36432$ $\quad$ $506 \times 9 \times 8 = 506 \times 72 = 36432$

Multiplication and Division of numbers

To multiply and divide numbers work from left to right

$309 \times 81 \div 9 = 25029 \div 9 = 2781$ $\quad$ $409 \times 81 \div 9 = 33129 \div 9 = 3681$

Division and Multiplication of numbers

To divide and multiply numbers work from left to right

$81 \div 9 \times 309 = 9 \times 309 = 2781$ $\quad$ $81 \div 9 \times 409 = 9 \times 409 = 3681$

Division of numbers

To divide numbers, work from left to right

$64000 \div 80 \div 8 = 800 \div 8 = 100$ $\quad$ $81000 \div 90 \div 9 = 900 \div 9 = 100$

Addition and Multiplication with Parentheses

Always first, work in parentheses

$(17 + 19) + 34 = 36 + 34 = 70$ $\quad$ $43 + (18 + 19) = 43 + 37 = 80$

$(25 + 5) \times 8 = 30 \times 8 = 240$ $\quad$ $6 \times (125+25) = 6 \times 150 = 900$

Addition and Subtraction with Parentheses

Always first, work in parentheses

$(700+300) -300=1000-300=700$ $\quad$ $(700-300) + 300 =400+300 =700$

$9000-(500+500) =9000-1000=8000$ $\quad$ $(1200-500) -500=700-500=200$

Multiplication and Division with Parentheses

Always first, work in parentheses

$706 \times (36\div4) =706\times9=6354$ $\quad$ $24028 \div (16\div4) =24028\div4 =6007$

Expressions with Parentheses:

Always first, work in parentheses

$15 \times (807-200) =15\times607=9105$ $\quad$ $(9000+105) \div15 =9105\div15=607$

$(17472-9000) \div 12=8472\div12=706$ $\quad$ $84072\div(10+ 2)=84072\div12=7006$

Do multiplication or division before addition or subtraction

Grade 6 Chapter 5 Summary 1

Division Numbers 4

Fourth operation with numbers is division ÷ or :

Division means separated number by some equal groups

First number in division called the dividend 6
Second number in division called the divisor 2
Result of division called the quotient 3
Dividend ÷ divisor = quotient $6 \div 2 = 3$

Properties of Division

Any number divided by one is the same number $5 \div 1 = 5$
Any number divided by the itself is one $5 \div 5 = 1$
Zero divided by any number is zero $0 \div 5 = 0$
Division by zero is not allowed $5 \div 0$ $A \div 0$ **none**

Formulas: 1) $A \div 1 = A$ 2) $A \div A = 1$
3) $0 \div A = 0$ 4) $A \div 0$ **none**

To check multiplication use division
$2 \times 3 = 6$ Check $6 \div 2 = 3$ or $6 \div 3 = 2$
To check division use multiplication and division
$6 \div 2 = 3$ Check $2 \times 3 = 6$ or $6 \div 3 = 2$

Division without remainder R = 0

```
            5 - Quotient
Divisor 7 | 35 – Dividend
           -35
             0 – Remainder
```

Division with remainder R > 0

```
            5 - Quotient
Divisor 7 | 36 – Dividend
           -35
             1 – Remainder
```

Grade 6 Chapter 5 Summary 2

Multiplication Numbers 5

Factor × Factor = Product (P) $2 \times 3 = 6$ (P)

Example 1

Top 3 2	1) 5 × 4 = 20	7) 5 × 7 = 35
6704	2) Put 0 below 5 (P)	8) Put 5 ahead 2 (P)
× 5	3) Put 2 above 0 (top)	9) Put 3 above 6 (top)
33520 (P)	4) 5 × 0 = 0	10) 5 × 6 = 30
	5) 2 + 0 = 2	11) 30 + 3 = 33
	6) Put 2 ahead 0 (P)	12) Put 33 ahead 5 (P)

Example 2

Top 2 3	1) 5 × 6 = 30	7) 5 × 4 = 20
7406	2) Put 0 below 5 (P)	8) Put 0 ahead 3 (P)
× 5	3) Put 3 above 0 (top)	9) Put 2 above 7 (top)
37030 (P)	4) 5 × 0 = 0	10) 5 × 7 = 35
	5) 3 + 0 = 3	11) 35 + 2 = 37
	6) Put 3 ahead 0 (P)	12) Put 37 ahead 0 (P)

Example 3

Top 2 3	1) 5×6=30	7) 5×4=20
8406	2) Put 0 below 5 (P)	8) Put 0 in ahead 3 (P)
× 5	3) Put 3 above 0 (top)	9) Put 2 above 8 (top)
42030 (P)	4) 5×0=0	10) 5×8=40
	5) 3+0=3	11) 40+2=42
	6) Put 3 ahead 0 (P)	12) Put 42 ahead 0 (P)

Example 4

Top 2 3	1) 5 × 6 = 30	7) 5 × 4 = 20
9406	2) Put 0 below 5 (P)	8) Put 0 ahead 3 (P)
× 5	3) Put 3 above 0 (top)	9) Put 2 above 9 (top)
47030 (P)	4) 5 × 0 = 0	10) 5 × 9 = 45
	5) 3 + 0 = 3	11) 45 + 2 = 47
	6) Put 3 ahead 0 (P)	12) Put 47 ahead 0 (P)

Grade 6 Chapter 5 Summary 2

Multiplication Numbers 6

Example 5

2
1
803
× 75
4015
+56210
60225

1) 75 = 70 + 5

2)
1
803
× 5
4015

3)
2
803
× 70
56210

Example 6

3
2
904
× 85
4520
+72320
76840

1) 85 = 80 + 5

2)
2
904
× 5
4520

3)
3
904
× 80
72320

Example 7

405
×106
2430 (P1)
+40500 (P2)
42930 (P)

1) 106 = 100 + 6

2) 405
× 6
2430 (P1)

3) 405
×100
40500 (P2)

Example 8

704
×205
3520
+140800
144320

1) 205 = 200 + 5

2) 704
× 5
3520

3) 704
×200
140800

Grade 6 Chapter 5 Summary 2
Division Numbers 7

2540 – dividend　　5 – divisor　　508 – quotient

Example 1　　**Every time bring down just one digit**

$\underline{\ \ 508}$	1) $5 \times \underline{5} = 25$	Put 5 in quotient
$5 \overline{)2540}$	2) $25 - 25 = 0$	
$\underline{-25}$	3) Bring down the 4	
04	4) $5 \times \underline{0} = 0$	Put 0 in quotient
$\underline{-0}$	5) $4 - 0 = 4$	
40	6) Bring down the 0	
$\underline{-40}$	7) $5 \times \underline{8} = 40$	Put 8 in quotient
0	8) $40 - 40 = 0$	

Example 2　　**3726 – dividend　　54 – divisor　　69 quotient**

Every time bring down just one digit

$37 < 54$　　$54 \times 6 = 324$　　$54 \times 7 = 378$　　$378 > 372$

$\underline{\ \ 69}$	1) $54 \times 6 = 324$	or	$3726 \,\lfloor\, \underline{54}$
$54 \overline{)3726}$	2) $372 - 324 = 48$		$\underline{-324}\ \ 69$
$\underline{-324}$	3) Bring down the 6		486
486	4) $54 \times 9 = 486$		$\underline{-486}$
$\underline{-486}$	5) $486 - 486 = 0$		0
0			

Example 3　　**Every time bring down just one digit**

$\underline{\ \ 7608}$	1) $5 \times \underline{7} = 35$	Put 7 in quotient
$5 \overline{)38040}$	2) $38 - 35 = 3$	
$\underline{-35}$	3) Bring down the 0	
30	4) $5 \times \underline{6} = 30$	Put 6 in quotient
$\underline{-30}$	5) $30 - 30 = 0$	
04	6) Bring down the 4	
$\underline{-0}$	7) $5 \times \underline{0} = 0$	Put 0 in quotient
40	8) $4 - 0 = 4$	
$\underline{-40}$	9) Bring down the 0	
0	10) $5 \times \underline{8} = 40$	Put 8 in quotient
	11) $40 - 40 = 0$	

Grade 6 Chapter 5 Summary 2
Division Numbers 8

Example 1

Every time bring down just one digit

```
     507
86 | 43602
    -430
      60
      -0
     602
    -602
       0
```

1) $86 \times \underline{5} = 430$ Put 5 in quotient
2) $436 - 430 = 6$
3) Bring down the 0
4) $86 \times \underline{0} = 0$ Put 0 in quotient
5) $60 - 0 = 60$
6) Bring down the 2
7) $86 \times \underline{7} = 602$ Put 7 in quotient
8) $602 - 602 = 0$

Example 2

6÷2=3 dividend 6 divisors 2 quotients 3(q)

Every time bring down just one digit

```
       708 (q)
905 | 640740
     -6335
       7240
      -7240
```

1) $905 \times \underline{7} = 6335$ Put 7 in quotient (q)
2) $6407 - 6335 = 72$
3) Bring down the 4
4) $905 \times \underline{0} = 0$ Put 0 in quotient (q)
5) $724 - 0 = 724$
6) Bring down the 0
7) $905 \times \underline{8} = 7240$ Put 8 in quotient (q)
8) $7240 - 7240 = 0$

Example 3

Every time bring down just one digit

```
       807 (q)
506 | 408342
     -4048
       3542
      -3542
          0
```

1) $506 \times \underline{8} = 4048$ Put 8 in quotient (q)
2) $451 - 448 = 3$
3) Bring down the 4
4) $506 \times \underline{0} = 0$ Put 0 in quotient (q)
5) $354 - 0 = 354$
6) Bring down the 2
7) $506 \times \underline{7} = 3542$ Put 7 in quotient (q)
8) $3542 - 3542 = 0$

Grade 6 Chapter 5 Summary 3

Order of Operations with Numbers 9

Multiplication and Addition of numbers

To multiply and add numbers always first multiply then add

Example 1 $5006 \times 5 + 4970 = 30000$

1) $5006 \times 5 = 25030$

2) $25030 + 4970 = 30000$

Example 2 $1960 + 6005 \times 8 = 50000$

1) $6005 \times 8 = 48040$

2) $48040 + 1960 = 50000$

Multiplication and Subtraction of numbers

To multiply and subtract numbers always first multiply then subtract

Example 3 $506 \times 80 - 30480 = 10000$

1) $506 \times 80 = 40480$

2) $40480 - 30480 = 10000$

Example 4 $55600 - 608 \times 75 = 10000$

1) 608×75: 3040, 42560, 45600

2) $55600 - 45600 = 10000$

Grade 6 Chapter 5 Summary 3

Order of Operations with Numbers 10

Division and Addition numbers

To divide and add numbers always first divide then add

Example 5

```
   2       1
7696+11520÷5=10000

      2304
1)  5)11520      2) 7696
   -10             +2304
     15            10000
    -15
      20
     -20
```

Division and Subtraction numbers

To divide and subtract numbers, always first divide then subtract

Example 6

```
   2       1
907 - 45192 ÷ 56 = 100
          807
1)  56)45192    2) 907
       -448       -807
         392       100
        -392
```

Example 7

6 ÷ 2 = 3 dividend 6 divisor 2 quotient 3

Every time bring down just one digit

```
907-307142÷506=907-607=300
1) 506×6=3036               607
2) 3071-3036=35     506)307142
3) 506×7=3542          -3036
                         3542
                        -3542
```

Grade 6 Chapter 3 Summary 3
Order of Operations with Numbers 11
Multiplication Numbers
To multiply numbers, work in any order

Example 1 $405 \times 8 \times 9 = 3240 \times 9 = 29160$

2) $\begin{array}{r} 405 \\ \underline{\times 8} \\ 3240 \end{array}$ 2) $\begin{array}{r} 3240 \\ \underline{\times 9} \\ 29160 \end{array}$

Example 2 $405 \times 8 \times 9 = 405 \times 72 = 29160$

2) $\begin{array}{r} 405 \\ \underline{\times 72} \\ 810 \\ \underline{28350} \\ 29160 \end{array}$

Multiplication and Division Numbers
To multiply and divide numbers work from left to right

Example 3 $309 \times 81 \div 9 = 25029 \div 9 = 2781$

1) $\begin{array}{r} 309 \\ \underline{\times 81} \\ 309 \\ \underline{+24720} \\ 25029 \end{array}$

2) $\begin{array}{r} 3681 \\ 9\overline{)25029} \\ \underline{-18} \\ 70 \\ \underline{-63} \\ 72 \\ \underline{-72} \\ 9 \\ \underline{-9} \end{array}$

Example 4 $208 \times 24 \div 6 = 832$

1) $\begin{array}{r} 208 \\ \underline{\times 24} \\ 832 \\ \underline{+416} \\ 4992 \end{array}$

2) $\begin{array}{r} 832 \\ 6\overline{)4992} \\ \underline{-48} \\ 19 \\ \underline{-18} \\ 12 \\ \underline{-12} \end{array}$

Grade 6 Chapter 5 Summary 4
Order of Operations with Numbers 12

Multiplication and Addition of numbers

To multiply and add numbers always first multiply then add

Example 5

$$5006 \times 5 + 4970 = 30000$$

1) $\begin{array}{r} 5006 \\ \times \underline{\quad 5} \\ 25030 \end{array}$ 2) $\begin{array}{r} 25030 \\ \underline{+ 4970} \\ 30000 \end{array}$

Example 6

$$1960 + 6005 \times 8 = 50000$$

1) $\begin{array}{r} 6005 \\ \times \underline{\quad 8} \\ 48040 \end{array}$ 2) $\begin{array}{r} 48040 \\ \underline{+1960} \\ 50000 \end{array}$

Multiplication and Subtraction of numbers

To multiply and subtract numbers always first multiply then subtract

Example 7

$$506 \times 80 - 30480 = 10000$$

1) $\begin{array}{r} 506 \\ \underline{\times\ 80} \\ 40480 \end{array}$ 2) $\begin{array}{r} 40480 \\ \underline{- 30480} \\ 10000 \end{array}$

Example 8

$$55600 - 608 \times 75 = 10000$$

1) $\begin{array}{r} 608 \\ \underline{\times 75} \\ 3040 \\ +\underline{42560} \\ 45600 \end{array}$ 2) $\begin{array}{r} 55600 \\ \underline{- 45600} \\ 10000 \end{array}$

Grade 6 Chapter 5 Summary 4
Order of Operations with Numbers 13
Division and Addition of numbers

To divide and add numbers always first divide then add

Example 1 $193230 \div 95 + 7966 = 10000$

1)
$$\begin{array}{r} 2034 \\ 95\overline{)193230} \\ -190 \\ \hline 323 \\ -285 \\ \hline 380 \\ -380 \\ \hline \end{array}$$

2) 7966 +2034=10000

Example 2 $7696 + 218880 \div 95 = 10000$

1)
$$\begin{array}{r} 2304 \\ 95\overline{)218880} \\ -190 \\ \hline 288 \\ -285 \\ \hline 380 \\ -380 \\ \hline \end{array}$$

2) 7696 +2304=10000

Division and Subtraction of numbers

To divide and subtract numbers always first divide after then subtract

Example 3 $462060 \div 765 - 104 = 500$

1)
$$\begin{array}{r} 604 \\ 765\overline{)462060} \\ -4590 \\ \hline 3060 \\ -3060 \\ \hline \end{array}$$

2) 604-104=500

Example 4 $1307 - 462060 \div 765 = 500$

1)
$$\begin{array}{r} 807 \\ 706\overline{)569742} \\ -5648 \\ \hline 4942 \\ -4942 \\ \hline \end{array}$$

2) 1307-807=500

Grade 6 Chapter 5 Summary 4
Order of Operations with Numbers 14

Expressions with Parentheses
Always first, work in parentheses

Example 5 $15 \times (807 - 200) = 15 \times 607 = 9105$

2) 807 − 200 = 607   2) 607 × 15: 3035 + 6070 = 9105

$$\begin{array}{r} 807 \\ \underline{-200} \\ 607 \end{array} \qquad \begin{array}{r} 607 \\ \underline{\times 15} \\ 3035 \\ \underline{+6070} \\ 9105 \end{array}$$

Example 6 $(200 + 105) \times 15 = 305 \times 15 = 4575$

1), 2)

$$\begin{array}{r} 200 \\ \underline{+105} \\ 305 \end{array} \qquad \begin{array}{r} 305 \\ \underline{\times 15} \\ 1525 \\ \underline{3050} \\ 4575 \end{array}$$

Example 7 $(17472 - 9000) \div 12 = 8472 \div 12 = 706$

1) 17472-9000=8472   2)

$$\begin{array}{r} 706 \\ 12\overline{)8472} \\ \underline{-84} \\ 72 \\ \underline{-72} \end{array}$$

Example 8 $84072 \div (10 + 2) = 84072 \div 12 = 7006$

1) $10 + 2 = 12$ 2)

$$\begin{array}{r} 7006 \\ 12\overline{)84072} \\ \underline{-84} \\ 072 \\ \underline{-072} \end{array}$$

Grade 6 Chapter 5 Summary 5
Table 1

Solve examples yourself then check your solutions

Examples 1

Multiply Numbers

1) 1056	2) 2056	3) 3506	4) 4506	5) 5506
× 5	× 5	× 5	× 5	× 5
30	30	30	30	30
250	250	2500	2500	2500
+5000	+10000	+15000	+20000	+ 25000
5280	10280	17530	22530	27530

6) 6704	7) 7406	8) 8406	9) 9406	10) 9046
× 5	× 5	× 5	× 5	× 5
20	30	30	30	30
3500	2000	2000	2000	200
+30000	+35000	+40000	+45000	+45000
33520	37030	42030	47030	47230

11) 104	12) 204	13) 304	14) 404	15) 504
×15	×15	×15	×15	×15
520	1020	1520	2020	2520
1040	2040	3040	4040	5040
1560	3060	4560	6060	7560

16) 604	17) 704	18) 804	19) 904	20) 25
×25	×25	×25	×25	× 904
3020	3520	4020	4520	100
+12080	+14080	+16080	+ 18080	+22500
15100	17600	20100	22600	22600

Grade 6 Chapter 5 Summary 5

Table 2

Solve examples yourself then check your solutions

Examples 2

Multiply Numbers

1) $\begin{array}{r} 12001 \\ \times\quad 9 \\ \hline 108009 \end{array}$ 2) $\begin{array}{r} 13002 \\ \times\quad 8 \\ \hline 104016 \end{array}$ 3) $\begin{array}{r} 23004 \\ \times\quad 7 \\ \hline 161028 \end{array}$ 4) $\begin{array}{r} 34005 \\ \times\quad 6 \\ \hline 204030 \end{array}$ 5) $\begin{array}{r} 56007 \\ \times\quad 5 \\ \hline 280035 \end{array}$

6) $\begin{array}{r} 67004 \\ \times\quad 9 \\ \hline 603036 \end{array}$ 7) $\begin{array}{r} 34005 \\ \times\quad 8 \\ \hline 272040 \end{array}$ 8) $\begin{array}{r} 45006 \\ \times\quad 7 \\ \hline 315042 \end{array}$ 9) $\begin{array}{r} 56007 \\ \times\quad 6 \\ \hline 336042 \end{array}$ 10) $\begin{array}{r} 67008 \\ \times\quad 5 \\ \hline 335040 \end{array}$

11) $\begin{array}{r} 604 \\ \times 205 \\ \hline 3020 \\ +120800 \\ \hline 123820 \end{array}$ 12) $\begin{array}{r} 704 \\ \times 205 \\ \hline 3520 \\ +140800 \\ \hline 144320 \end{array}$ 13) $\begin{array}{r} 804 \\ \times 205 \\ \hline 4020 \\ +160800 \\ \hline 164820 \end{array}$ 14) $\begin{array}{r} 904 \\ \times 205 \\ \hline 4520 \\ +180800 \\ \hline 185320 \end{array}$ 15) $\begin{array}{r} 205 \\ \times 904 \\ \hline 820 \\ +184500 \\ \hline 185320 \end{array}$

16) $\begin{array}{r} 1200 \\ \times 15 \\ \hline 6000 \\ +12000 \\ \hline 18000 \end{array}$ 17) $\begin{array}{r} 2304 \\ \times 25 \\ \hline 11520 \\ +46080 \\ \hline 57600 \end{array}$ 18) $\begin{array}{r} 3405 \\ \times 35 \\ \hline 17025 \\ +102150 \\ \hline 119175 \end{array}$ 19) $\begin{array}{r} 4506 \\ \times 45 \\ \hline 22530 \\ +180240 \\ \hline 202770 \end{array}$ 20) $\begin{array}{r} 5607 \\ \times 65 \\ \hline 28035 \\ +336420 \\ \hline 364455 \end{array}$

Grade 6 Chapter 5 Summary 5

Table 3

Addition and Subtraction
Rules and Examples

Addition

First number in addition called the addend	2	$2 < 5$	
Second number in addition called the addend	3	$3 < 5$	
Result of addition called the sum	5	$5 > 2$	$5 > 3$

Addend + Addend = Sum $2 + 3 = 5$

First Addend = Sum – Second Addend

$? + 3 = 5$ $? = 5 - 3$ $? = 2$

Second Addend = Sum – First Addend

$2 + ? = 5$ $? = 5 - 2$ $? = 3$

Subtraction

First number in subtraction called the minuend	5	$5 > 2$	$5 > 3$
Second number in subtraction called the subtrahend	2	$2 < 5$	
Result of multiplication called the product	3	$3 < 5$	

Minuend – Subtrahend = Difference $5 - 2 = 3$

Minuend = Difference + Subtrahend

$? - 2 = 3$ $? = 3 + 2$ $? = 5$

Subtrahend = Minuend – Difference

$5 - ? = 3$ $? = 5 - 3$ $? = 2$

Grade 6 Chapter 5 Summary 5

Table 4

Multiplication and Division
Rules and Examples

Multiplication

First number in multiplication called the factor	2	$2 < 6$	
Second number in multiplication called the factor	3	$3 < 6$	
Result of multiplication called the product	6	$6 > 2$	$6 > 3$

Factor × Factor = Product $2 \times 3 = 6$

First Factor = Product ÷ Second Factor

$? \times 3 = 6$ $? = 6 \div 3$ $? = 2$

Second Factor = Product ÷ First Factor

$2 \times ? = 6$ $? = 6 \div 2$ $? = 3$

Division

First number in division is called dividend	6	$6 > 2$	$6 > 3$
Second number in division is called divisor	2	$2 < 6$	
Result of division is called quotient	3	$3 < 6$	

Dividend ÷ Divisor = Quotient $6 \div 2 = 3$

Divisor = Dividend ÷ Quotient

$6 \div ? = 3$ $? = 6 \div 3$ $? = 2$

Dividend = Quotient × Divisor

$? \div 2 = 3$ $? = 3 \times 2$ $? = 6$

Students!! Attention!!

Adults, help please students 1 – 4 grades to read and understand advice.

When I was in 5th grade I did not like math. I am not always understood math. I was 11 years girl. Next 6th grade I got a very good teacher. She explains very clearly and I understood her very well. I was worked very hard and I was not afraid to ask questions. Final Grade in Algebra in 6th grade I got an A. I said in class that I want to be a teacher of math. But some students said that I am not enough smart to be a math teacher. I said I will be work hard with math. I did it, I got a Master's degree in math and education, I love my students, I was teaching many years, in High School, part-time college math instructor, full-time university math lecturer.

Some people do a simple job, eat, sleep, go to work. People with a bad brain have not so a good job, not so good salary. If you look at TV how someone's dance, you will be not changing your brain. But if you take a dancing class this helps your body and brain ability. The same if you want to use the calculator, but the calculator not always with you. Much better if you learn how you need to do any operations with numbers without a calculator. If you read this math book you will know what you need to do. This teaches you how to improve your memory, attention, logic thinking, how to read your math books every day, your brain works better. You can work 50 minutes with math, 10 minutes break maybe dance. Do not afraid to do mistakes, fixed and move to the next step. Don't listen to anyone, choose what do you like more.

Students need to know if you work with the brain every day, your brain works better. You need a good brain in any job, in any class. When you finish High School, you need to get any profession, you need to go to college and get a good job and a better salary. You need to learn any subject, its work for the brain. Read the book, do the exercises yourself, try your best, and then check your answer at the end of the chapter. Be honest with yourself. You need to believe in yourself. The brain is like muscles. People with good brain lives longer. The professors at my university were over 80 years old, wise, and optimistic. It was a great pleasure to talk to them.

Made in United States
Cleveland, OH
05 November 2026

47669889R00121